中学HighSchool/北京 BEIJING/王昀 wangyun/2003

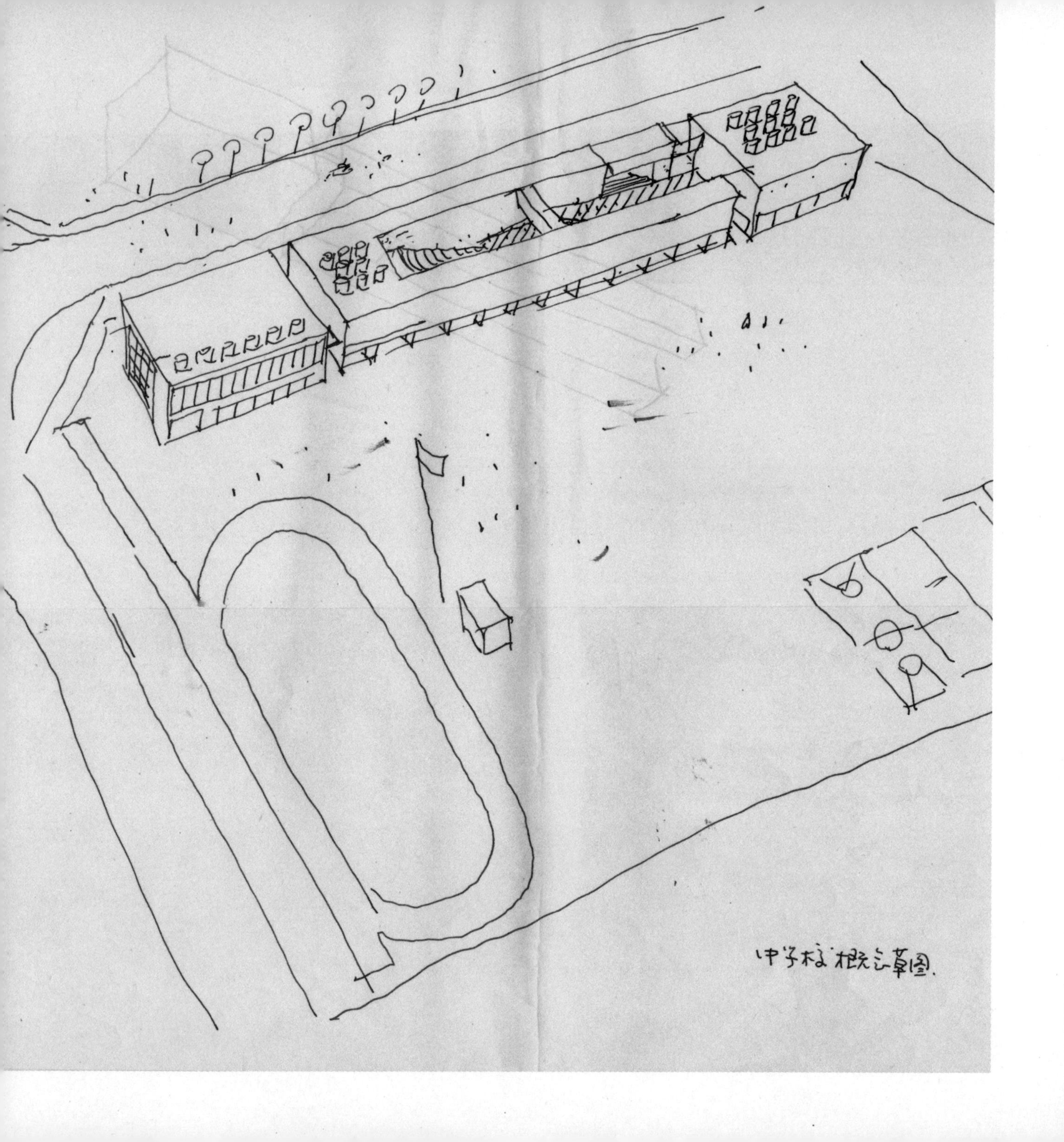
中学校概念草图.

中学的概念构思草图
Concept sketch

# 中学 High School

中国若尔盖地区的红光聚落呈“一”字形排开，形成一个绵延1公里的线形聚落预示着百子湾中学的意象
1-km-long linear China Zoige area Hongguang settlement indicates Baiziwan High School's concept

“一”字形的校舍全长约157米，是一个为60万平方米住宅社区服务的24班中学校舍

In-line 24-class schoolhouse are about 157 meters long, serving for a 600000$m^2$ residential community

N

0 5 20m

## 首层平面图 / The 1st floor plan

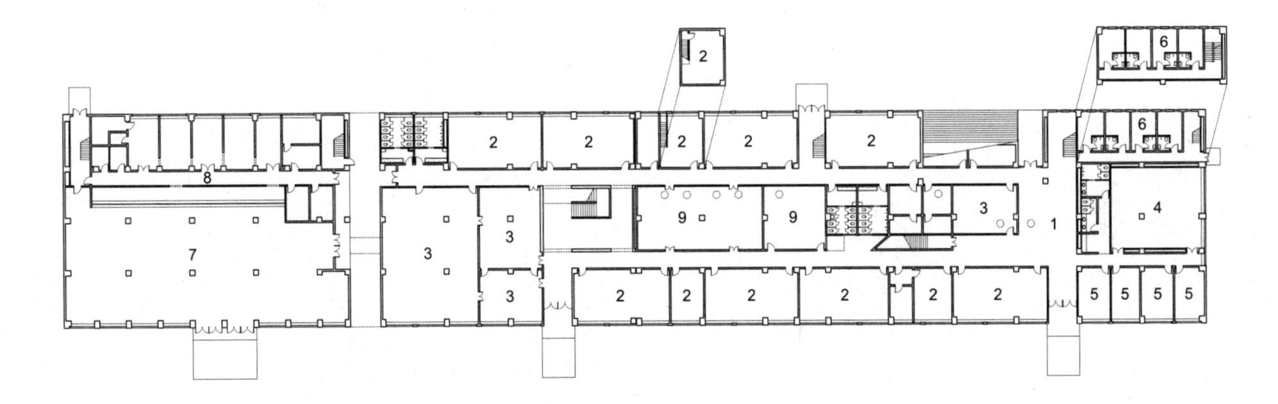


1 门厅 Hall
2 教学用房 Teaching room
3 阅览室 Reading room
4 总务仓库 Logistics warehouse
5 办公室 Office
6 教师宿舍 Teacher's dormitories
7 餐厅 Dining room
8 厨房 Kitchen
9 展览空间 Exhibition space

| | | |
|---|---|---|
| 1 | 教学用房 | Teaching room |
| 2 | 办公室 | Office |
| 3 | 教师宿舍 | Teacher's dormitories |
| 4 | 普通教室 | Common classroom |
| 5 | 风雨操场 | Open playground |
| 6 | 室内休息区 | Indoor relaxation area |
| 7 | 室外活动平台 | Outdoor activity platform |

二层平面图 / The 2nd floor plan

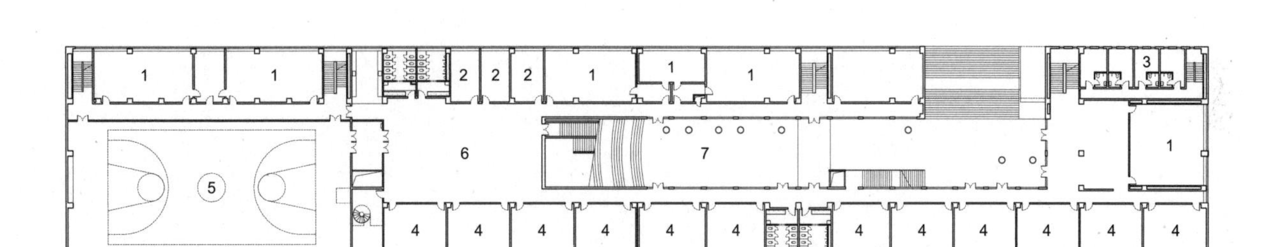

三层平面图 / The 3rd floor plan

1 教学用房 Teaching room
2 办公室 Office
3 普通教室 Common classroom
4 室内休息区 Indoor relaxation area
5 阶梯教室 Terrace classroom

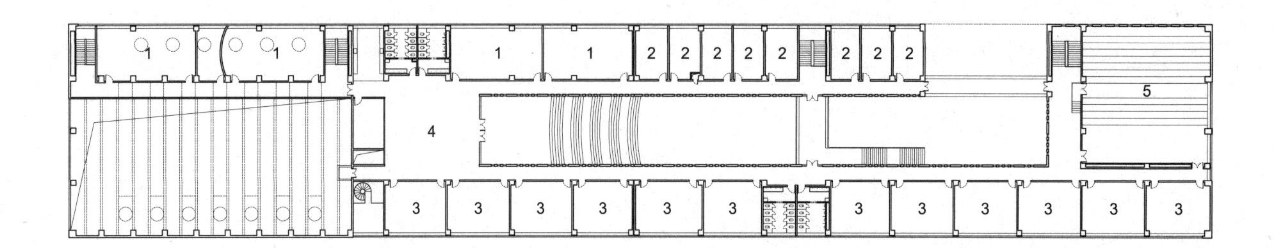


学校二层的中庭有通往三层的大台阶和楼梯，使得整个中庭成为学校内部的内院广场，在这里可进行集会活动。学校主入口上空有一个跨度为16米的空中连桥，使三层的交通形成一个环路

The big stairways leading to the third makes the floor 2nd-floor atrium the interior square to hold gatherings. A 16-meter-span overpass on the top of school entrance forms a circuit for the third floor.

在二楼中庭，学生们从各方向汇聚进行交流、玩耍和休息而不受干扰
Students come to the atrium to play and take a break from various directions without inter-fering each other

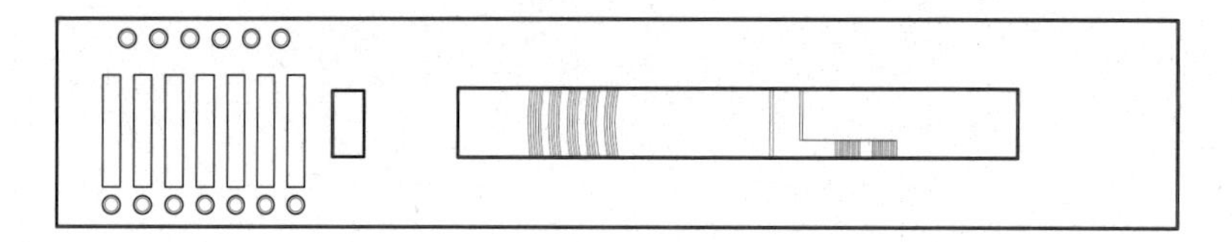

# 王昀

# 西溪学社

西溪湿地

第三期

艺术村

**WANG YUN**

# XIXI

**INSTITUTE**

**2007**

# 8

## 西溪学社
## Xixi Institute

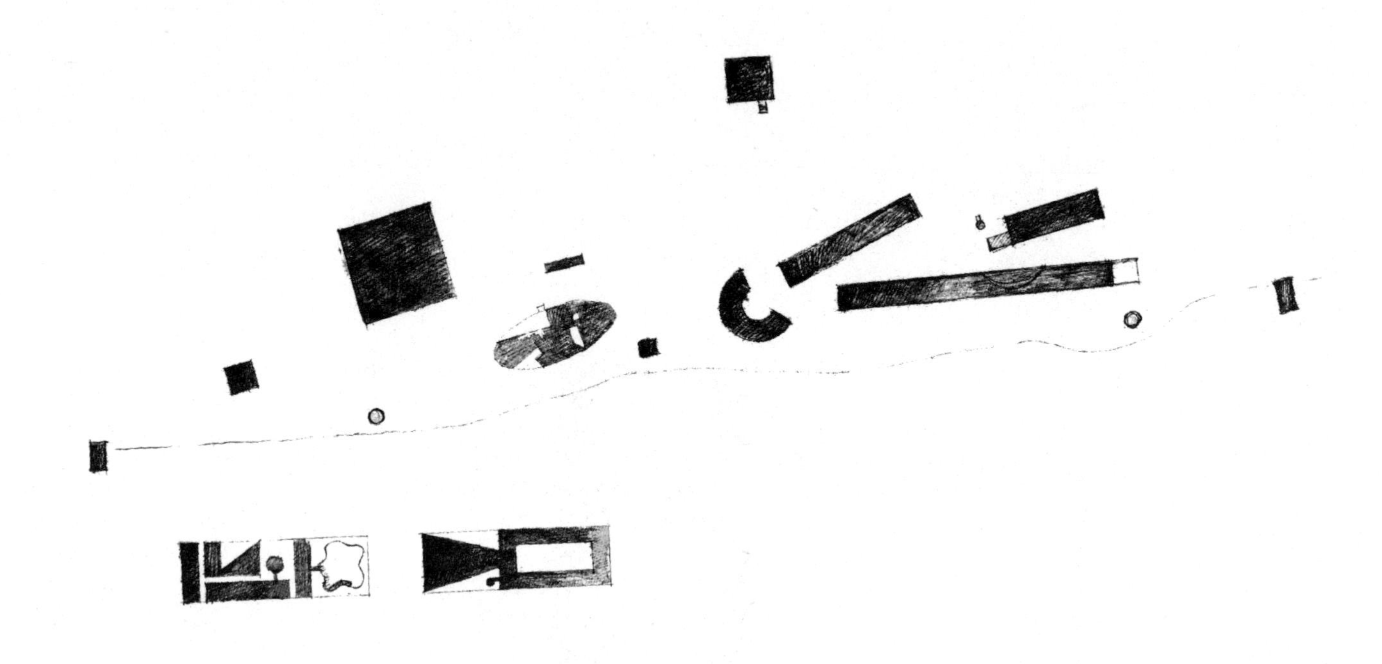

西溪学社建筑群的构思草图
Concept Sketches of Xixi Institute Complex

作为一组群像而存在的湿地建筑
Group-portrait Architectures on the Wetlands

西溪学社是西溪湿地第三期艺术村中12个项目中的一个局部项目。业主希望为文化和艺术界人士提供一个进行艺术创作和学术交流的场所。西溪学社建设用地的现状分散且彼此被水塘所分隔。面对这样的用地状况，如何满足既有集中，又有分散的功能需求是设计的关键所在。考虑到西溪湿地中树木的尺度，设计时将建筑体量拆分，采用聚落的离散布局方式是设计的主旨。西溪学社共由10个建筑和6个构筑物组成，建筑采用明确的几何形体组合，以使其产生共同幻想特征。同时注重了聚落中所必需的“微差”设置，使体验者能够在若即若离中感受到一种丰富性的存在。在中国传统的绘画中，留白是非常重要的。受这种在白纸上以墨作画的启发，将建筑做成如白色的宣纸，作为背景衬托湿地的自然景观，让建筑作为背景融于画中，与湿地的自然环境互相映衬。

Xixi Institute is one of 12 Xixi Wetland Art Village Phase Ⅲ projects. The client wanted to provide a meeting place for culture exchange and art creation. Xixi Institute's site is scattered and separated by ponds. How to fulfill the functions of concentration and disper-sion is the key to design. Considering the scale of Xixi Wetland trees, the design disconnects the architectural volume and adopts a dispersed settlement layout. Xixi Institute is com-posed of 10 buildings and 6 constructions. It's a combination of pure geometric blocks, which evokes collective fantasy. Meanwhile, the subtle variance is attended to, so people can feel the richness in the dreamlike environment. Blank-leaving is vital. Inspired by ink wash painting on white paper, I designed the building as white Xuan paper to set off the wetland scape. The natural foreground perfectly blends into the architectural background.

西溪学社建筑群的聚落状态
Settlement Pattern of Xixi Insititute Complex

西溪学社建筑群平面图
Plan of Xixi Insititute Complex

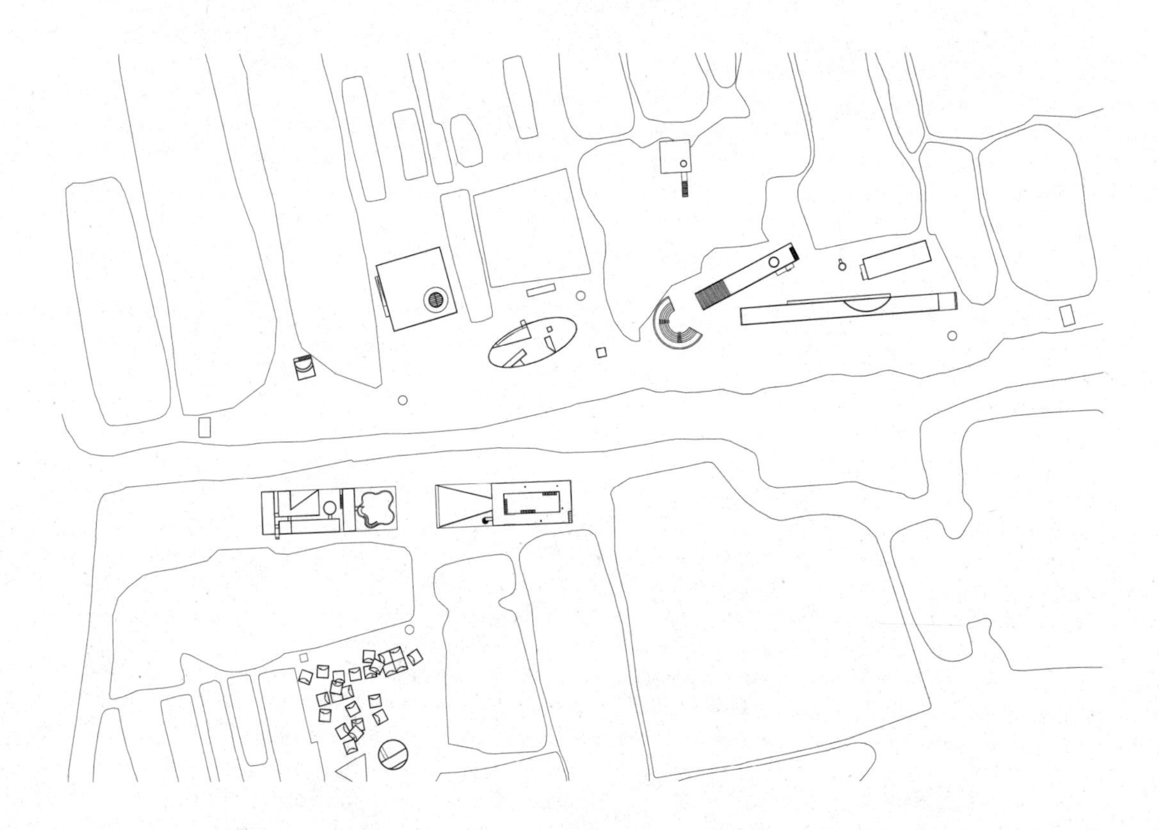

# 幼儿园

王昀

# Kindergarten

wangyun

2003

# 幼儿园
Kindergarten

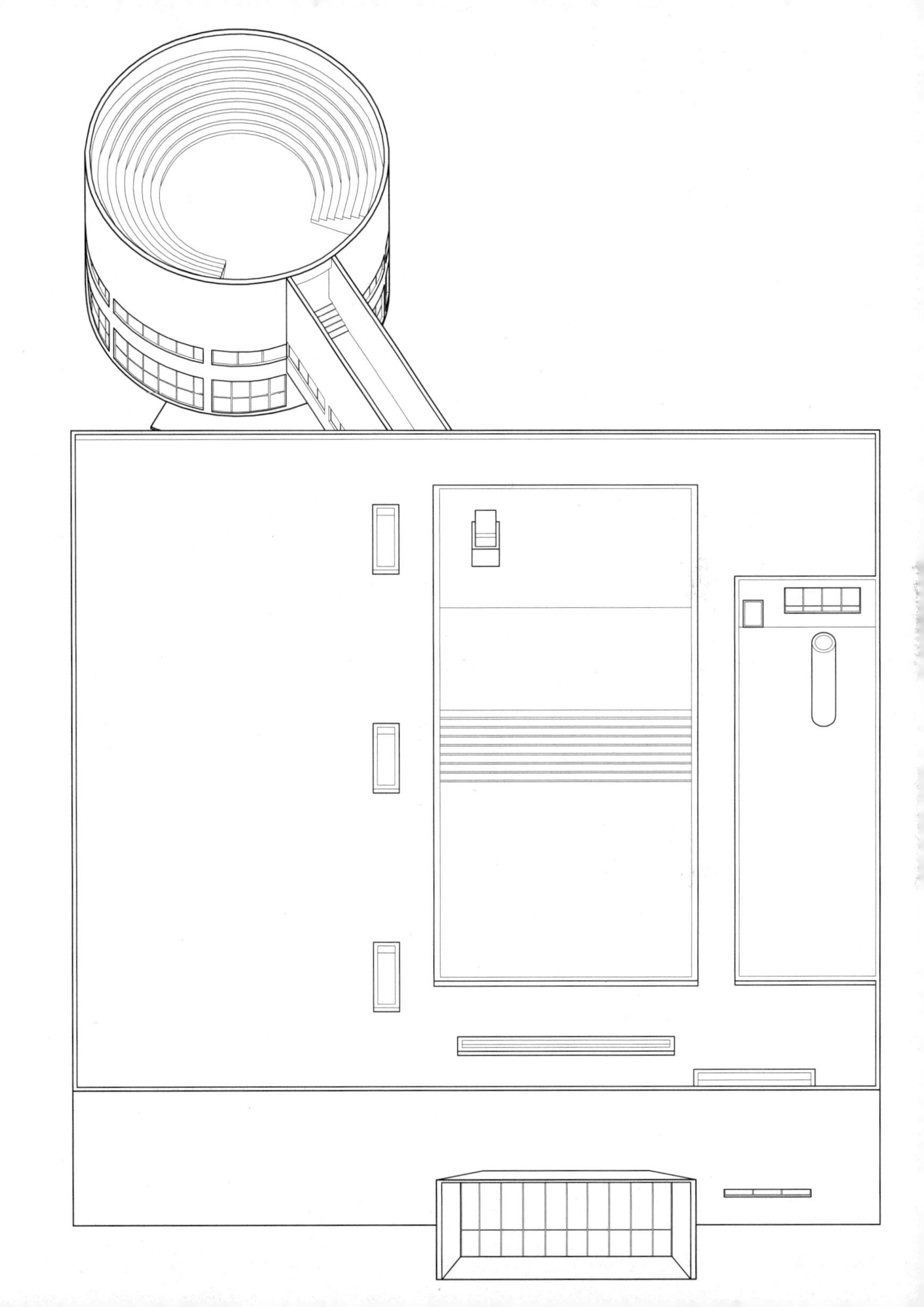

轴测图 / Axonometric drawing

北京百子湾幼儿园占地面积4000平方米，总班级数为9个。该幼儿园总建筑面积3200平方米，为三层。整个建筑由一长36.2米、宽37.5米、高10.5米的方形体块与一个半径7米、高10.5米的圆柱体块组合而成。作为建筑主体的方形体块共三层，进入喇叭形入口，是一个三层高的大厅，大厅的上空设有一个窄长条的天窗将阳光导入，伴随时间的不同，条形的光线能够在中庭的墙面上形成不同的光影交融。入口大厅的左侧是教室区域，右侧是厨房和办公区。在幼儿园的南侧区域集中布置活动室和孩子们的卧室。北侧则在一层布置厨房和办公，二层布置一个图书室。三层布置有一个露天的活动场地。体块的中心设有一个内院，从而使得方形体块形成一个“虚中”的意向并与窑洞建筑相互印合。

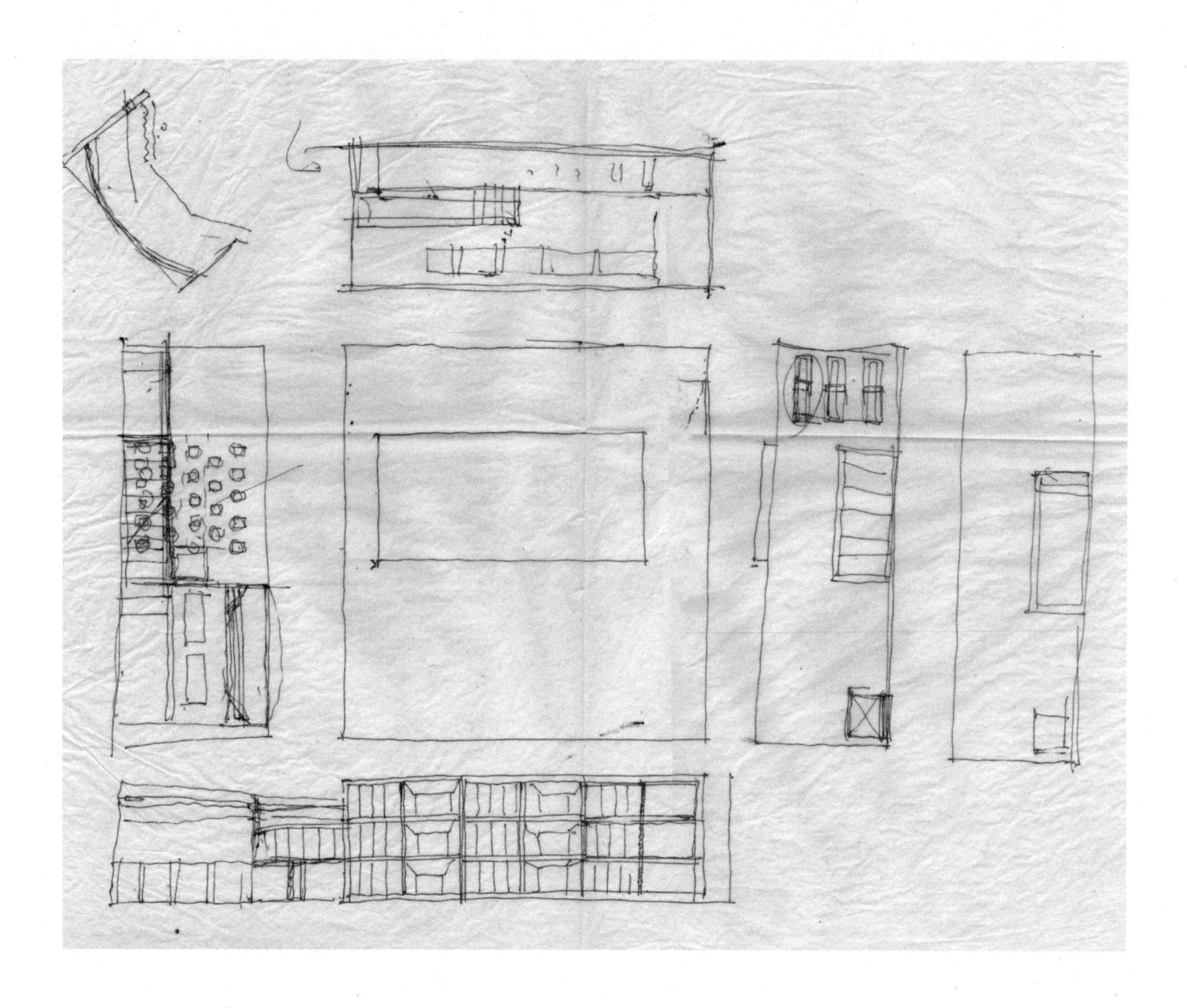

Beijing Baiziwan Kindergarten has 9 classes; gross site area is 4000m$^2$. The three-floor kindergarten's building area is 3200m$^2$, composed of a 36.2mx37.5mx10.5m cubicle and a 7-meter-radius 10.5-meter-tall cylinder. The cubicle as the main part has three floors. Entering the trumpet there is a three-floor-tall entrance hall. The long and narrow skylight on the ceiling introduces sunshine into the hall. The stripe light will cast different shadows as time changes. Classrooms are on the left of entrance hall; kitchen and offices are on the right; recreation room and children's bedrooms are in the south part of the kindergarten. For the north part, kitchen and offices are on the first floor, library is on the second floor, and an open-air play ground is on the third floor. There is a courtyard in the center, thus make the hollow cubicle echo with cave dwellings.

首层平面图 / The 1st floor plan

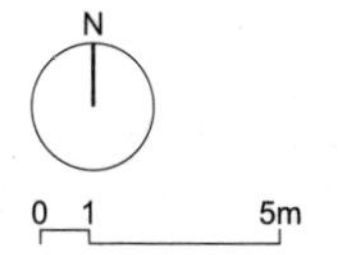


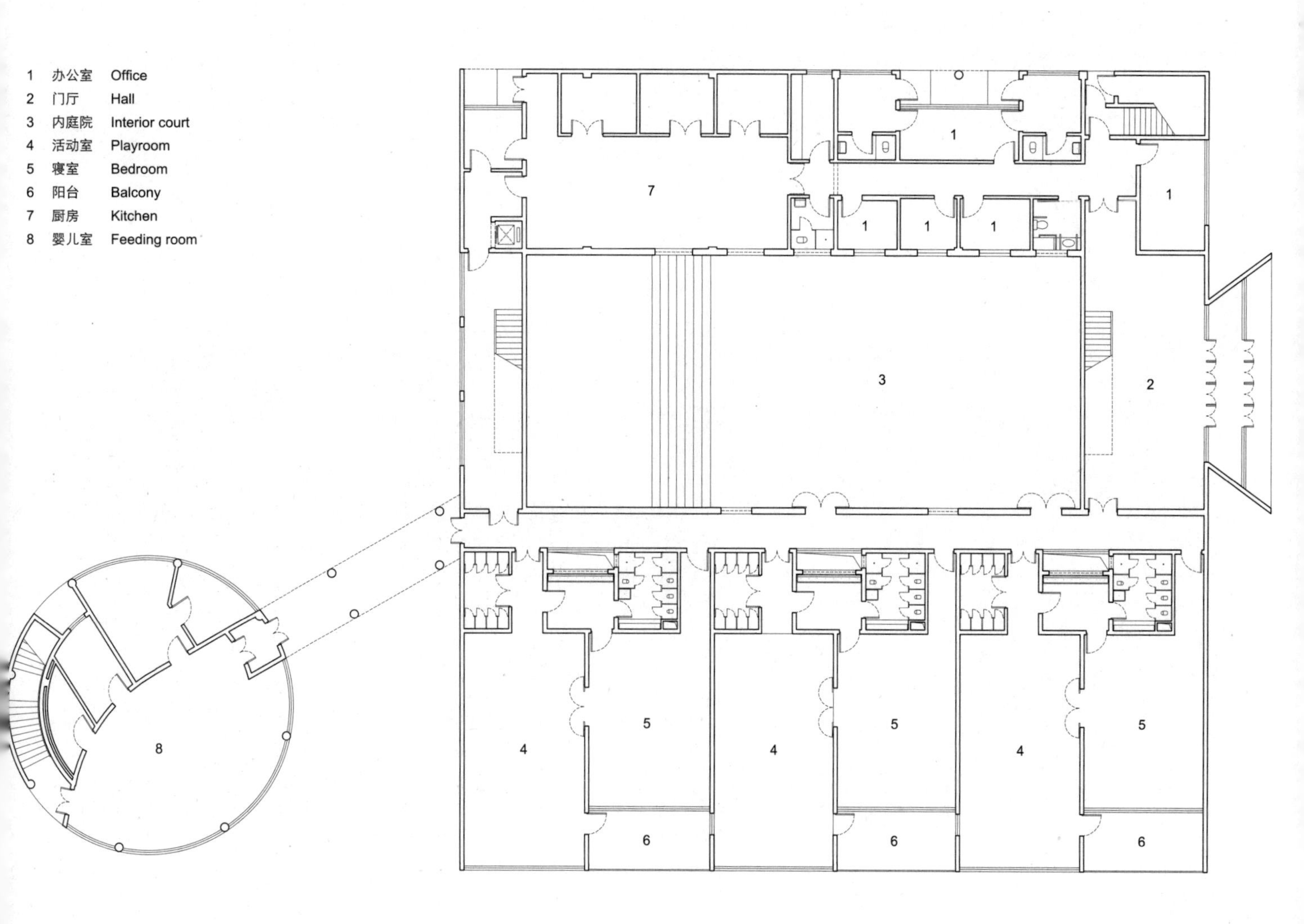

太平岭吴姓居民的窑洞住居中间是一个长方形的院子，院的四周墙壁侧向挖有拱形横向洞穴作为居室使用。窑洞东侧有一个坡道是窑洞的主入口。

There is a rectangular courtyard in the centre of Taipingling Wu Family's cave dwelling. The arch caves as bedrooms are vertical to surrounding walls of the courtyard. A ramp to the east of cave is the main entrance.

1 办公室 Office
2 活动室 Playroom
3 寝室 Bedroom
4 阳台 Balcony
5 阅览室 Reading room
6 音体教室 Music classroom

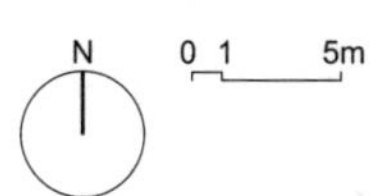


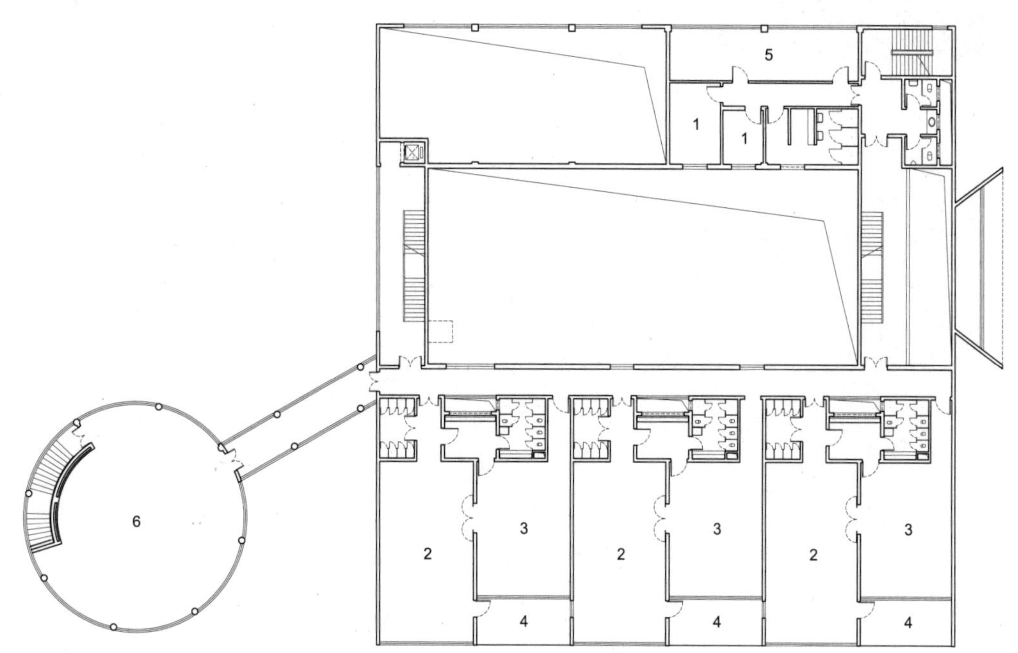


二层平面图 / The 2nd floor plan

中国青海藏族的帐篷住居的内部 Inside of Qinghai Tibetan tent dwelling

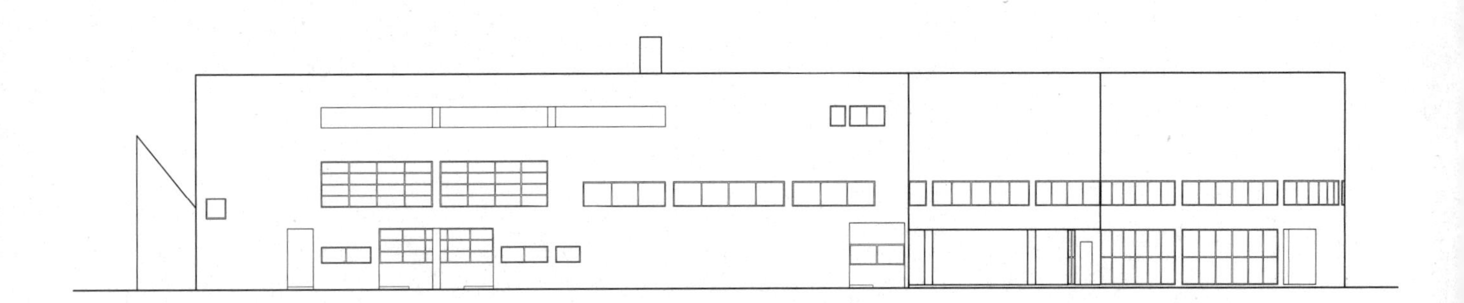

北立面 / North elevation

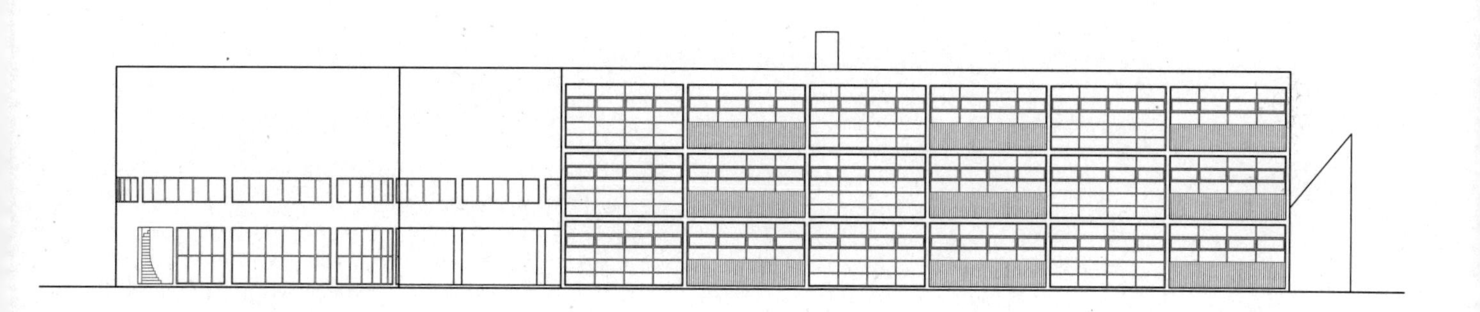

南立面 / South elevation

屋顶平面图 / Roof floor plan

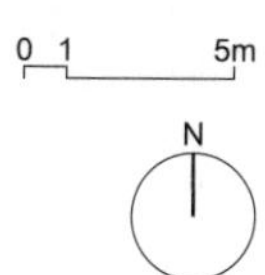


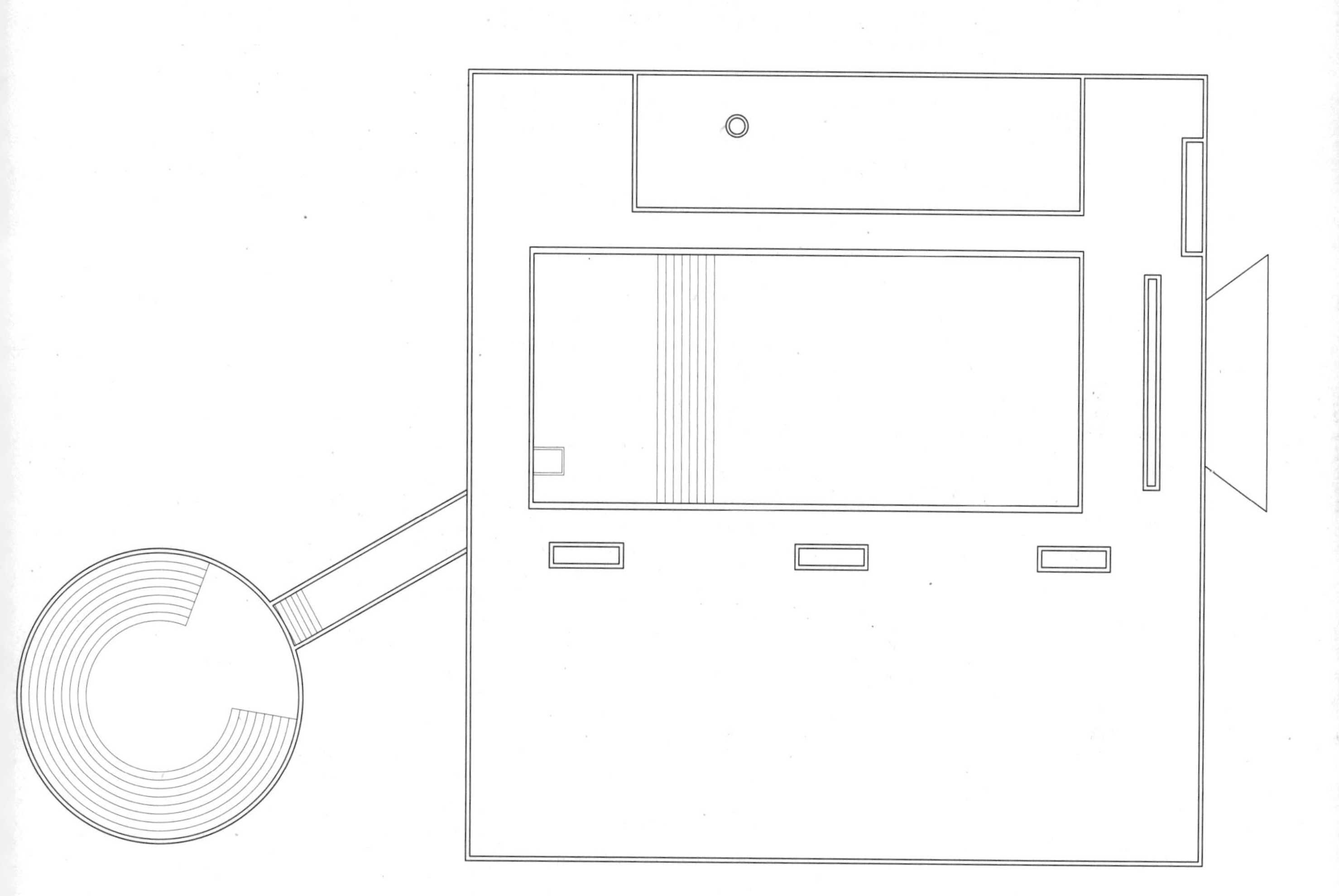

# 空间的陈述

# Overview of Eight Projects

Wang Yun

王昀

中国建筑工业出版社

# 8空间的陈述/王昀
# Overview of Eight Projects/Wang Yun

图书在版编目（CIP）数据

8空间的陈述 / 王昀. – 北京 ： 中国建筑工业出版社，
2015.5
ISBN 978-7-112-17929-9

Ⅰ. ①8… Ⅱ. ①王… Ⅲ. ①建筑设计—作品集—中国—
现代 Ⅳ. ①TU206

中国版本图书馆CIP数据核字（2015）第053630号

责任编辑：徐　冉
责任校对：姜小莲 关 健
封面设计：宁　晶
版式设计：赵冠男

感谢北京建筑大学建筑设计艺术研究中心建设项目的支持

**8空间的陈述**
Overview of Eight Projects
**王昀**
Wang Yun
*
中国建筑工业出版社出版、发行（北京西郊百万庄）
各地新华书店、建筑书店经销
北京顺诚彩色印刷有限公司印刷
*
开本：889×1194毫米 1/20 印张：9 字数：242千字
2015年6月第一版 2015年6月第一次印刷
定价：52.00 元
ISBN 978 - 7 - 112 - 17929 - 9
(27140)

## 序

在过去10年的工作中，我经手了近百个项目，但绝大多数在设计过程中便先后夭折，有的是因业主对设计“风格”有所微辞，有的则是在过程中由于“种种原因”发生变化，有的甚至已经开工建设，但由于“原因种种”而最终变更了设计。这几本小册子呈现给大家的是在这样的现实中得以生存下来的，可谓生命力极强的8个设计。将这些设计的思考过程总结并呈现给各位读者，是对其瞬间绽放的青春般生命力的记录和纪念，因为这些小册子所呈现的绝大多数设计事实上已经消失或发生了改变。

## Preface

Over the last decade, I've dealt with around 100 projects, yet most of them came to an untimely end. Some were due to clients disagreeing with the design 'style'; some encountered unexpected changes for 'various reasons'; and some were altered out of 'different causes' even when construction already took place. These pamphlets present 8 vital projects which were able to survive such situations. Development processes are dedicated to readers as a record or in memory of their momentary vitality, since most of them have disappeared or been modified.

# 目录
# Contents

## 设计意图
Concept of Design

1  60平方米极小城市 60m² Mini City

2  门与厅 Door and Hall

3  财政局 Finance Bureau

4  会所 Clubhouse

5  住宅A + B Houses A+B

6  幼儿园 Kindergarten

7 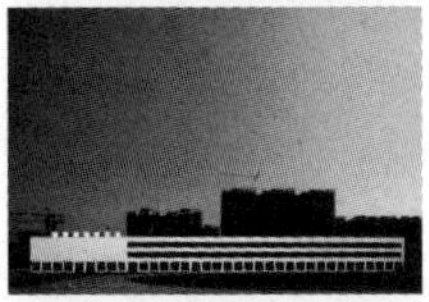 中学 High School

8  西溪学社 Xixi Institute

# 60平米极小城市 $60m^2$ Mini City

# 1

这是一个6层的砖混结构住宅，三室一厅，南侧有两个居室，北侧有一个居室，中间是起居室。开始看这个房子的时候，感觉一个大的问题，就是卫生间的门是朝向客厅的。虽然房子中该有的设施都具备，但从人的生活行为来分析，处处充满弊端。如何让这个住宅变得有生活感觉，是设计面临的一个主要问题。住宅不仅仅是一个面积大小的问题，让空间变得有趣，生活起来感觉到空间呈现不断扩大的状态，其实是这个设计的初衷。在画平面图的时候，一些带有记忆性的空间在设计过程当中会呈现于笔端。场景的不断产生，同时又结合场景进行布局，功能本身随之附加进去。或许设计就是这样的一个糅合的过程。在这个过程中，思考最多的是卫生间的门的问题。最终采用的方式是将卫生间的门组合到梯段状的壁柜上面，表面上看，卫生间的门成为壁柜上的组合门从而起到了隐身的效果。为了增加空间的扩大感觉，尽可能地展示建筑的物理距离是一个有效的方法。实际上这个住宅的最长距离是从北向南的两个居室的贯通处，有十几米的距离，利用这个距离设置一个狭长的街道让视线有足够的穿越，从而会获得通透的满足。同时视线在空间里游走的过程中，长短之间的对比，对应地会在居者心里产生微妙的感受。此外设计时如何让室内空间产生回路是扩大空间感的重要手段。因为一般，室内空间的围合有赖于四周的墙壁，但是人的视线在巡视的过程中，四壁造成的封闭性造成了室内空间的封闭感觉，设计时在室内的中间放了一个柜子，将原有空间分为大小两个空间，空间自己便循环起来，在从大空间到小空间的往复过程中，获得空间的自我比对关系。

It's a 3-bedroom flat on the 3rd floor of a 6-floor brick-concrete building. Two bedrooms in the south, one in the north, and living room in the middle. Major problem is bathroom opens to living room. Although basic facilities are equipped, there are lots of behavioral defects. How to make this flat lively is the designer's top priority. It's not only about area, but also about fun. I hope to make the residents feel increasing space when actually living here. Some space memories haunted me during plan design. Scenes kept emerging, combined with layout and function design. Maybe design is such a mixture progress. I spent most of the time figuring out a solution to bathroom's door. In the end, I integrated the door to cas-cade-shape closet, which looks like a whole set. Show the architecture's physical distance to a large extent, is a method to space increase. In fact, the longest distance in the flat is over 10 meters - from north bedroom to south bedroom. Make full use of this distance to build a narrow passage enables satisfaction of people seeing all the way through. In the meantime, contrast between long and short in the architectural promenade initiates subtle emotions. What's more, design a circuit is an important approach to enlarge space. Normally, indoor enclosure depends on the surrounding walls. But, in this case, when people wander around, the walls create closeness. So I set a cabinet in the center, dividing the space into two parts, one large and one small, which lead to spontaneous circulation. Self-contrast is gained be-tween the large space and the small one.

门与厅是一个80多平方米的办公楼入口设计。基地在考察时还是一个非常杂乱的环境，当时场地周围盖了很多小的库房，南侧是一栋普通的砖混结构住宅楼，北侧是一栋有着米黄色水刷石表面的住宅楼，远处还有一栋写字楼，是以咖啡色调为主。原有办公楼本身也是一栋米黄色建筑。在都是暖色调的空间中，整体感觉有些燥气。我想如果让人们的视觉中心都能够落到新加建的门厅上面或许可以改变周边这种夺人视线的杂乱环境，而这本身是设计的一个初衷。

从摩洛哥马拉喀什通往(Taroudant)路边的住居，白色涂料泼洒式地喷涂在建筑的关键位置转移了原有建筑的视觉中心，门与厅的设计从这里开始。设计试图采用强烈的黑白对比，让现有的所有杂乱成分变成“中间调子”，在环境中注入新的黑白灰关系，使得整个调子重新产生阶梯关系并彼此协调起来。简而言之，就是加一个最重颜色的同时再加一个最浅颜色。设计时，我们将原有的米黄色办公楼刷成黑色，然后把入口做成一个白色的体块，概念设计就此完成。

摆放在广场上的白色体块，根据场地布局，微微地弯曲做成弧形体块，考虑的是从侧面进入入口时能够让人产生一种亲切的迎合感。同时在入口处的视觉中心设计了一个小的花坛，夏天花草从中蔓出，并成为第一个视觉中心，伴随着人的慢慢走进，当整个门厅的建筑全都露出来时，圆形的花坛和门厅之间产生一刚一柔的对比。此外，微微弯曲的门厅内部还会造成人在透视上的错觉，造成一种不聚焦的透视学错乱。

It's an $80m^2$ entrance design. The site used to be in disorder: many small warehouses surrounding the courtyard; an ordinary brick-concrete building in the south, a beige granite plaster building in the north; a brown office building in the distance; and the original office building is beige as well. All warm colors create some kind of anxiety. My primary intention was to draw people's attention to the new foyer to reduce the environment influence.

Houses are alongside the road from Morocco Marrakech to Taroudant. White paint splashes on the major part of buildings, which transfers the original visual center. This is where Door and Hall design starts. The contrast between intense black and white attempts to turn all the disorder components into grey. the new black-white-grey relation gradually reconciles the environ-ment's color tone. in short, adds a darkest color and a lightest color at the same time. We painted the original beige office building into black, and designed the entrance new building as white block; thus accomplished the concept design.

White blocks on the square are slightly curved to suit the plot layout, welcoming the people coming in from side entrance. A small flowerbed is set in the entrance visual center. in the summer time, full blossom becomes the first visual center. While approaching until the whole foyer building fully exposed, there is hard and soft contrast between round flowerbed and foyer. Meanwhile, curved indoor space creates defocusing perspective illu-sion.

## 财政局 Finance Bureau

石景山财政局项目用地的面宽非常狭窄，其西侧是一个已经完工的法院建筑。使用方对于设计有一个要求，希望办公楼将对外的出入口放在南侧，对内的出入口放在北侧。南侧这个出入口主要是满足财政局对外办公的需要，要求设置一个大堂。建筑在一层设接待大厅，二层有一个大台阶可以起到分流的作用。同时入口处的大台阶希望能够成为早晚为普通市民开放的一个公共性广场。这个办公楼由40米×40米×60米的体块构成。南侧巨大的门是面对城市呈现开放姿态的大剧场，为周边居民提供了一个能够在傍晚乘凉的场所。室内采用了L形的布局形态，因为北京西面还是比较晒的，所以办公空间基本设置在东侧和南侧，迎着西侧这部分是一个采光筒，能让光一束一束地从侧墙和空中洒进，形成一个阳光大厅。而这个大厅实际上还有另一作用，就是让所有办公空间的门都面向这个大厅，当办公人员出入办公空间的瞬间，能够体会到空间在视线上的一种变化，同时在心情上获得一种转换。一层有一个大的报告厅，报告厅的屋顶布置成为一个室内的屋顶花园。大厅的西侧和屋顶，共排列有28个直径为3米的圆窗，体现“凿户牖以为室”的空间观念。“凿”这个概念是埋藏在中国人心底的一种解决空间问题的观念，同时也是表示一个动作。我们曾经调查过的古崖居聚落、窑洞聚落，无一不体现这种“凿”的意向。也是因为北京延庆的古崖居聚落，启发我们在石景山“凿”出一个财政局。这个建筑还有一个大的技术难点，就是前面的这个门框的正面只有5毫米。

Shijingshan Finance Bureau is located on a very narrow plot, with a newly completed court building to its west. The client requires public entrance to be on the south side, and staff entrance on the north. South public entrance needs a business hall on the first floor. Big stairs leading to the second floor can effectively divert the flow. And hopefully it can become a public square sooner or later. The building is in a 40m×40m×60m block. The huge door on the south is an embracing theater providing a leisure space. Indoor adopts an L-shape. Since the west rooms get very hot in the afternoon, office space is designed in the east and south side. A lighting cylinder is set in the west wall to enable daylight to come in through side walls and ceiling, therefore forms a sunlight atrium. All the offices open to the atrium, so the staff will experience different moods when they pass through the sunlight atrium. There is a big lecture hall on the first floor, with an indoor roof garden on the top. Twenty-eight 3-meter-diameter round windows in the west side and on top of the atrium represent traditional concept of "chisel the door and window to make a room". "Chisel" is believed to be the architectural solution by Chinese, and also stands for an action. The ancient cliff settlement and cave dwelling we surveyed both embody this Chisel concept. And, it's also because of this Beijing Yanking cliff settlement that inspires us to chisel a Finance Bureau in Shi-jingshan. There is another technique problem - the front door frame is only 5mm wide?

庐师山庄会所用地的西侧有一片大的中心绿地，使会所成为能让使用者体验中心庭院的装置所在是设计的最初动机。使用者希望会所的一层为咖啡厅，二层为供居民们展示自己作品的画廊和举办一些交流性质活动的场所，三层设置两个大的多功能性会议室。设计时，考虑到西侧的风景，在一层大堂的一侧顺着景观的方向设计了一个长坡道，目的是让人不停留在一个视点去看一个对象物，而是通过漫步并不断升高的过程来体会窗外风景的变化。在二层一侧还有一个旋转楼梯通达三层。这个旋转楼梯不仅是将从一层到二层的坡道过程中所观察到的平面性风景，伴随这个通往三层的路径旋转进一步扭曲，其本身也构成了二楼的视觉趣味中心所在。这个旋转楼梯在施工的时候其实遇到了很多困难，曾经施工单位提出施工难度大，要求做一个轻质的钢梯（我们的施工单位，经常会找出理由尽可能地降低施工难度），但在我们的强力要求之下，最后施工方还是用混凝土把它打出来了。会所西侧的幕墙是一个难点。这个幕墙高11米，下面玻璃约3.6米高，上面的玻璃将近7米高，玻璃的宽度是1.8米。幕墙花费了长时间的论争，结果该幕墙不经意间挑战了我国制作幕墙的技术极限。克服了幕墙遇到的困难之后，入口这个3.6米×3.6米的大门成为厂家头痛问题。他们担心这个门将来有可能打不开。而结果出乎他们的意料，实际面临的是如何将门关上的问题。如何挑战或者做一些非日常性、非常规性的东西是需要勇气的，而让我们困扰的正是生产厂家似乎不太愿意去应对挑战。

There is a large green space in the west side of Lushi Clubhouse's site. The design's primary intention is to make the users experience the installation in the atrium. The client required a cafe on the first floor, a gallery to exhibit residents' works and hold some activities, two big multi-function meeting rooms on the third floor; and several small guest rooms on one side of the first floor. A long sightseeing ramp is designed along the west side of lobby. The architectural promenade enables people seeing different views outside the window while escalating. There is a spiral stairway from 2nd to 3rd floor. It twists the plane ramp view from 1st to 2nd floor, at the same time itself becomes the visual centre of 2nd floor. The spiral stairway creates many difficulties for the construction company, who proposed to change to a lightweight stairway. But we insisted on the original design, in the end they built it with concrete. Curtain wall on the west facade is a problem. It's 11 meters tall - 3.6m below and 7m above - and 1.8 meters wide. It took a long discussion to decide on curtain walls, which turned out to overcome China's technique limit. Afterwards, the 3.6m×3.6m entrance door became troubling. The manufacturers were afraid the door might not open properly, yet surprisingly, they had to deal with how to shut it. Make it challenging or create some irregularity requires some courage. Anyhow, most manufacturers are reluctant to take the challenge.

庐师山庄别墅区位于北京八大处工人疗养院西侧，原来是北京建工集团的疗养所，原来的基地中心有一个小的园林。由于用地紧张，设计时如何创造出属于住户自己的环境，同时又满足密度的要求，是一个难点。其实在中国传统的民宅中，这个问题早已有了好的解决办法，那就是采用内向式的布局方式，即在住宅的内部重新围合出一个私有的世界。在外面，房子和房子之间是紧挨着的，一旦你跨进门，那个天地就完全属于住户自己。住宅A+B，采用两个18米见方的联拼方式构成，18米×18米这个尺度，源于我曾经调查和测绘过的中国青海土族民居。土族人在这个18米见方的夯土墙里解决了自己的居住问题，这或许是一个非常适宜人支配的范围和领域。换句话来说，在这个范围内走来走去安排生活是适宜的。因为人们在建造时，房子究竟做多大，其实是按照一种非常长久的经验来进行判断的，18米×18米是经验积累的产物，也恰好是舒服和亲切的尺度。支配范围的尺度确定后，便开始内部的空间组织，从居住者开门想起，努力思考人在里面怎样展开生活，同时眼前是一个什么样的场景，一步一步地走进去，完成整体的空间体验，而不仅仅靠面积来对住宅进行支撑，完全依靠空间的运作。从最终的结果上来看，住宅A的空间感相对集中，很多的景象和思绪可能是在一个比较集中的状态下完成的，但在做住宅B的时候，感觉好像没有住宅A来得错动和复杂，显得比较轻松。其实住宅B在中间庭院有一个回路，这是我小时候到姑姑家，从她家前院到后院过程的路径在这里投射的结果。

Lushi Mountain Villa are located to the west of Beijing Badachu Workers' Sanitarium. It used to be Beijing Construction Engineering Group's sanitarium with a small garden in the center. Due to limited site area, how to create a private environment and meet density requirement at the same time, is a hard task. In fact, we have effective solution in traditional Chinese vernacular dwellings, which is to build a private world by means of enclosed layout. Houses are close to one another on the outside. Once you're inside, you own the whole world. Houses A+B are composed of two 18mx18mx18m blocks. The 18m module origi-nates from Qinghai Tu Nationality's vernacular dwellings I surveyed. They solve the residential problems within the 18mx18mx18m puddle walls. It might be a very appropriate area for people's activities. In other words, this is an approachable space to live. After all, how big the house is depends on longtime experience. Thus how the 18m comes about, and it's also a comfortable and friendly module. When the ballpark is confirmed, I started to think about indoor layout. Imagine how the residents actually live since they enter the door, one step after another, walk through every scene they encounter to complete a whole spatial experience. The design does not rely on floor area, but more of a spatial sequence. Seen from the final result, House A is more concentrated, while House B is more dispersed, which reflect the two mental states I was in when I designed - A intense and B relaxed. There is a circuit in House B's atrium. It is borrowed from my childhood experience to my aunt's home - route projection of front yard to back yard.

幼儿园 Kindergarten

幼儿园项目是2003年初开始设计的，是北京百子湾经济适用房小区的配套设施，设定为9个班的规模。设计采用简单的几何学体块组合，让儿童从小就对几何学形成多意的读解和想象。积木其实就是一个小尺度几何学物体，运用最简单的几何形体，通过拼接和拼搭，诱发孩子们的想象力和抽象的理解力。这个幼儿园的几何体构成关系，实际上不仅仅在于体块上的构成，更重要的是培养儿童去认识形体和空间之间的对应关系。儿童时期的经历和经验对于未来成长发挥着至关重要的作用。而建筑当中最重要的一点，就是空间这个东西如何让儿童能够获得感受，这一点实际上是这个设计当中很重要的一个问题。这个幼儿园在表面形体上，基本上是三块，一个是圆，一个是方，还有一个实际上是三角锥。三角锥插到方的里面后被切掉形成一个梯形，变成了一个小喇叭作为幼儿园入口，而这个小喇叭事实上又想唤起“小喇叭开始广播了”这一广播节目的联想。幼儿园中间挖出一个庭院，这个庭院事实上是窑洞的意象。走在窑洞的上面，俯瞰那个从地面上切下的完整和漂亮的方形空间，我想其与从周围围合的高层住宅上对其进行俯视的感觉是一样的。在庭院里也可以举办一些节目。这里设有一个舞台，通过几个小的踏步构成制造出一种舞台的气氛。幼儿园的圆形体块的屋顶上，还有一个露天小剧场，儿童可以在屋顶上唱歌，享受阳光。这个圆形剧场还有特别声效，因为圆形拢音，儿童在中间唱歌时，稍微发一点声就感觉声音特别饱满并因此增加自信。

It started since early 2003, as Beijing Baiziwan affordable housing compound's supporting facilities. The kindergarten has 9 classes. The simple geometric blocks combination enables the children gain first impression and vision of geometrics. Building blocks are actually small-scale geometric objects. Putting up simplest blocks evokes children's imagination and abstract comprehension, which is the kindergarten design's primary intention. Childhood experience plays an important role in one's adult life. Most important of my design is to influence the kids with architectural space. The kindergarten is composed of three blocks: one round, one cubicle, and one triangular pyramid. The triangular pyramid is cut off inside the cubicle and becomes a trapezoid, which is the trumpet kindergarten entrance. This trumpet also recalls the memory of a past broadcasting program - Little Trumpet is On Air. There is a courtyard in the centre of kindergarten as a metaphor for cave. Looking down at the carved underground square space from the surrounding high-rise apartment buildings feels the same as from above the cave. Meanwhile, the 'stage' made by several steps can host some shows. On the roof of kindergarten's round block is a small open-air theater where kids can sing on the rooftop in the sunshine. This circular theater also has special vigorous sound effects because of its shape, making the singing children feel very confident.

# 中学 High School

百子湾中学与百子湾幼儿园同在一个小区里，两个方案基本在同一个时间完成。因为周围都是二十几层的高层住宅，希望从上面向下看时像一个巨大的窑洞是这个设计的最初设想。在陕西我测绘过这样一个窑洞，在一片玉米地里，正好是一个中午去的，在上面行走时，你根本看不到房子的存在。只能听到从玉米地里传出收音机播出的"小说连续广播"，突然就在眼前出现一个大坑，大坑的里面是一个长条形窑洞，里面有小孩在戏耍。这是兄与弟两家共同挖的大窑洞，它不同于一般的方形窑洞，而是一个长条形的地下空间，所有人的生活都在大窑洞中展开，其容纳和承载力非常高。在设计学校的瞬间，这个兄弟俩的窑洞突然浮现于脑海。在设计时还有一个很重要的细节，与我上中学的经历有关。我的初中是在哈尔滨第五中学度过的，操场对面有一个公共卫生间，课间仅10分钟，下课以后，学生得先从楼上拥下，然后有的奔厕所，有的在操场上放放风，刚走到厕所，马上又打铃上课了，不得不往教室跑。怎么样在这个学校的设计当中避免此情景，也是一个重要的考量。在设计时，我尽可能将厕所分成端头一个，中间一个。还有就是二层的教室旁边做一个室外的大平台，因为我原来就读的中学是4层，三层、四层的学生要下到一层去活动，非常不便。所以我想怎么样让在上层教室的学生非常自然地、顺利地走到室外是设计中一个重要的落脚点。我们将所有的教室沿着中庭这面走廊直接开门。学生下课时可以直接到中庭活动，更接近活动场地，然后三层有一个大台阶，学生可以很顺当地走下来。

Baiziwan High School and Baiziwan kindergarten are within the same compound, and completed almost at the same time. They're surrounded by 20-floor high-rise apartment buildings so I hope the design seems like a huge cave looking down from above. I surveyed a cave in a cornfield in Shanxi Province. It was at noon then, the dwelling is invisible from walking above. You can only hear the Novel Serial broadcasting program from the radio in the cornfield, and all of a sudden, the big pit appeared in front of your eyes, with kids playing in a rectangular cave inside. It is a big rectangular cave dug by two brothers, different from usual square caves. Two families' live in this high-capacity cave. The scene hit me when I designed the high school. Another important detail is related to my junior high school experience at Harbin No. 5 High School. The public bathroom is across the playground. 10-min break is very tight for so many students to go down to the playground and then run to the bathroom. Sometimes the class began when students just got to the bathroom and had to run back. How to avoid such experience is my thinking. I tried my best to set the bathrooms in the centre and at one end. And an outdoor platform near the 2nd floor classrooms, because it was very inconvenient for 3rd and 4th floor students to go down to the ground to take a break at my old high school. Thus comes forth the outdoor space in the middle floor. All the classrooms open to the atrium corridor so the students can go directly to the atrium after class.

西溪学社 Xixi Institute

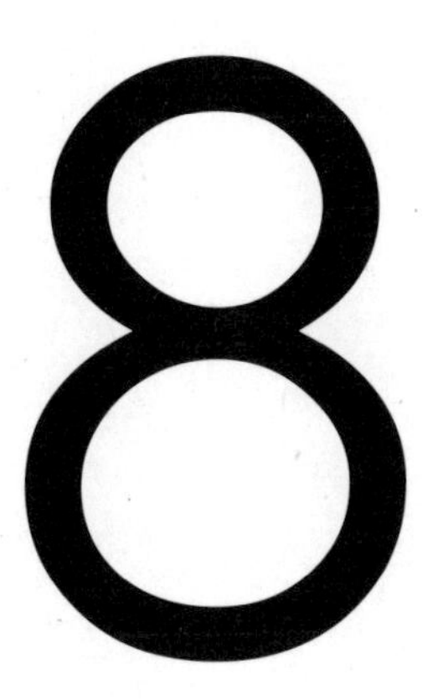

西溪学社项目是由一名叫“黄石”的小说家策划的，他对建筑有特定的情感追求。他参与了杭州西溪湿地第三期艺术村的策划工作，就决定召集十二个中国设计师进行一个集群设计。我被分配到这个H地块，去现场考察时，发现地块中可以盖建筑的部分非常分散，还有一个事实是这个地块里面树木的尺度是舒服和宜人的。面对环境，离散式聚落的布局方式显现出来，让每一个房子之间都有距离，每一个单体又具有相对独立性。这便是设计的开始。按照黄石的建议，这里希望能够做一个西溪学社。从文化的角度，杭州的西湖边上曾经有一个西泠印社，为中国的近代书法，特别是篆刻文化作出过非常大的贡献。黄石提倡西溪学社，是希望未来建筑师和艺术家可定期云集在这里，进行学术交流和探讨。这个想法甲方是赞同的。依据这样的意图，设计时安排有展厅、报告厅以及住宿的场所。如项目中的A1，设计了一个可提供二十四个独立居住单元的聚落式小宾馆。之所以称其为聚落式小宾馆，是因为这个宾馆的平面布局是直接引用我曾经测绘的中国青海日月山村，把日月山村的空间格局，通过尺度转换完整地变成一个宾馆的状态，希望把日月山村村民曾经有的集体记忆，通过我这个调查者的个人记忆和重新读解，在西溪湿地这个另外的地域上获得新的转换和生命的延续。我也希望在这个宾馆盖成之后来看看与现实当中日月山村之间在外部空间上的对应关系，究竟有哪些不同。此外，园区A5项目中收集光线的巨大高塔，与杭州的六和塔以及曾经调查过的希腊桑托里尼岛上的聚落相关联。

The project was planned by a novelist called Huang Shi, who has a specific emotional pursue for architecture. He participated in the project planning of Hangzhou Xixi wetland art village Phase Ⅲ and decided to call 12 Chinese architects to do a complex design. Plot H was assigned to me. I found architectural plot very scattered when i went on field research. Another fact to mention is that, the scale of trees on the plot is very pleasant. Adjust to the environment, dispersed settlement layout is adopted - every building is independent. Huang Shi suggested to build a Xixi Institute. Culturally speaking, Xiling Engraving Institute near Hangzhou West Lake is famous for modern Chinese calligraphy, especially engraving culture. Xixi Institute is supposed to regularly gather architects and artists to have academic exchange and discussion. Party A agrees to the idea. Therefore, the design includes show-room, lecture hall and guest rooms. A1 is a 24-room settlement hotel. I call it settlement because the layout is borrowed from Qinghai Riyueshan Village I once surveyed. The space structure is turned into a hotel via scale transition. This is a tribute to Riyueshan Villagers' collective memories through my personal experience and understanding. The lives are renewed and continued on Xixi wetland. I also hope to find out the difference and space relation between this hotel and Riyueshan Village. Besides, A5 is a tall light tower, related to Hangzhou Liuhe Pagoda and Greece Santorini settlement.

# 作者简介

王昀博士

1985年　毕业于北京建筑工程学院建筑系获学士学位

1995年　毕业于日本东京大学获得工学硕士学位

1999年　毕业于日本东京大学获得工学博士学位

2001年　执教于北京大学

2002年　成立方体空间工作室

2013年　创立北京建筑大学建筑设计艺术研究中心担任主任

2015年　于清华大学建筑学院担任设计导师

建筑设计竞赛获奖经历:

1993年日本《新建筑》第20回日新工业建筑设计竞赛获二等奖

1994年日本《新建筑》第4回S×L建筑设计竞赛获一等奖

主要建筑作品:

善美办公楼门厅增建，60m$^2$极小城市，石景山财政局培训中心，庐师山庄，百子湾中学，百子湾幼儿园，杭州西溪湿地艺术村H地块会所等

参加展览:

2004年6月参加"'状态'中国青年建筑师8人展"

2004年首届中国国际建筑艺术双年展参展

2006年第二届中国国际建筑艺术双年展参展

2009年参加在比利时布鲁塞尔举办的"'心造'——中国当代建筑前沿展"

2010年参加威尼斯建筑艺术双年展，德国Karlsruhe Chinese Regional Architectural Creation建筑展

2011年参加捷克布拉格中国当代建筑展，意大利罗马"向东方-中国建筑景观"展，中国深圳·香港城市建筑双城双年展

2012年第十三届威尼斯国际建筑艺术双年展中国馆等

# Author Biography

Dr. Wang Yun

Graduated with a Bachelor's degree from the Department of Architecture at the Beijing Institute of Architectural Engineering in 1985.

Received his Master's degree in Engineering Science from Tokyo University in 1995.

Received a Ph.D. from Tokyo University in 1999.

Taught at Peking University since 2001.

Founded the Atelier Fronti (www.fronti.cn) in 2002.

Established Graduate School of Architecture Design and Art of Beijing University of Civil Engineering and Architecture in 2013, served as dean.

Served as a design Instructor at School of Architecture, Tsinghua University in 2015.

Prize:

Received the second place prize in the "New Architecture" category at Japan's 20th annual International Architectural Design Competition in 1993

Awarded the first prize in the "New Architecture" category at Japan's 4th SxL International Architectural Design Competition in 1994

Prominent Works:

ShanMei Office Building Foyer, 60m$^2$ Mini City, the Shijingshan Bureau of Finance Training Center, Lushi Mountain Villa, Baiziwan Middle School, Baiziwan Kindergarten, and Block H of the Hangzhou Xixi Wetland Art Village.

Exhibitions:

The 2004 Chinese National Young Architects 8 Man Exhibition, the First China International Architecture Biennale, the Second China International Architecture Biennale in 2006, the "Heart-Made: Cutting-Edge of Chinese Contemporary Architecture" exhibit in Brussels in 2009, the 2010 Architectural Venice Biennale, the Karlsruhe Chinese Regional Architectural Creation exhibition in Germany, the Chinese Contemporary Architecture Exhibition in Prague in 2011, the "Towards the East: Chinese Landscape Architecture" exhibition in Rome, and the Hong Kong-Shenzhen Twin Cities Urban Planning Biennale

# 住宅A+B Houses A+B／王昀 WANG YUN 2003

# 住宅 Houses

A + B

摩洛哥塔真多托（Tazentout）聚落，住居是院落式中间围合形态，开孔的院落本身构成一幅“重复”风景而与中国的院落式住宅相印合。

Morroco Tazentout Settlement. Enclosed courtyard dwelling is like a repetitive view, resonating with Chinese courtyard houses.

位于北京西山八大处附近的庐师山庄别墅区是由52栋院落式小住宅所构成的，住宅A+B是其中最大的两栋住宅。每一栋住宅的地上和地下面积共约700平方米，且在住宅的东侧两栋分别各延伸出两个大小为18米×l2米的内院，内院中有分别有楼梯能够直通到住宅的地下室部分。两栋住宅是由两个长和宽均为18米、高为7米的方盒子拼合联立在一起的，并且均为地上二层地下一层。两个住宅的空间性格，一个刚性些，一个相对暧昧柔性。两个住宅在整体的设计上采用抽象的白色箱体进行空间构造。在内部空间组织上，设计者力图将其意识中的风景物象化。室内穿插游走的散步路径，步移景异的空间景象，以抽象的景观作用于体验者。眼前一幕幕抽象风景的呈现，在激唤起体验者自身经验记忆与联想的同时，使体验者获得高次元意识的直观。

The 52-courtyard Lushi Mountain Villas are located near Beijing Badachu in west suburb. Houses A+B are the biggest, each has a total floor area of 700m$^2$ both above and under the ground. To the east of each house attached an 18m×12m courtyard and stairways leading to underground rooms. Two houses are composed of two 18mx18mx7m boxes, two floors above ground and one floor under. The houses have different space characteristics: one masculine, and the other effeminacy. They both adopt white abstract boxes appearance and interior materialized views. The architectural promenade accommodates the passengers with abstract landscape, which evokes their own memories and gain direct higher dimen-sion experience.

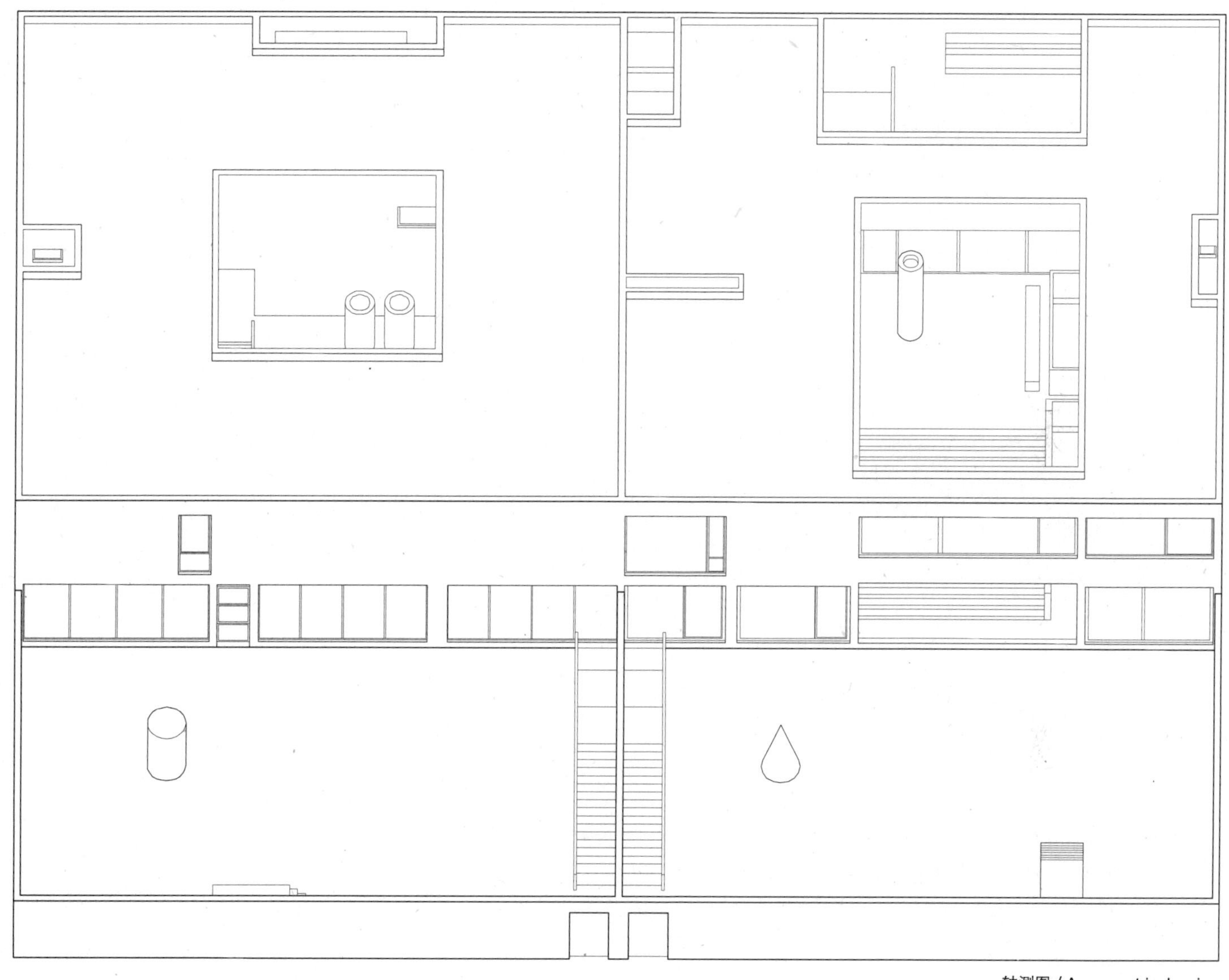

轴测图 / Axonometric drawing

左图：云南城子村的住居
Left Figure：Yunnan Chengzicun Residence

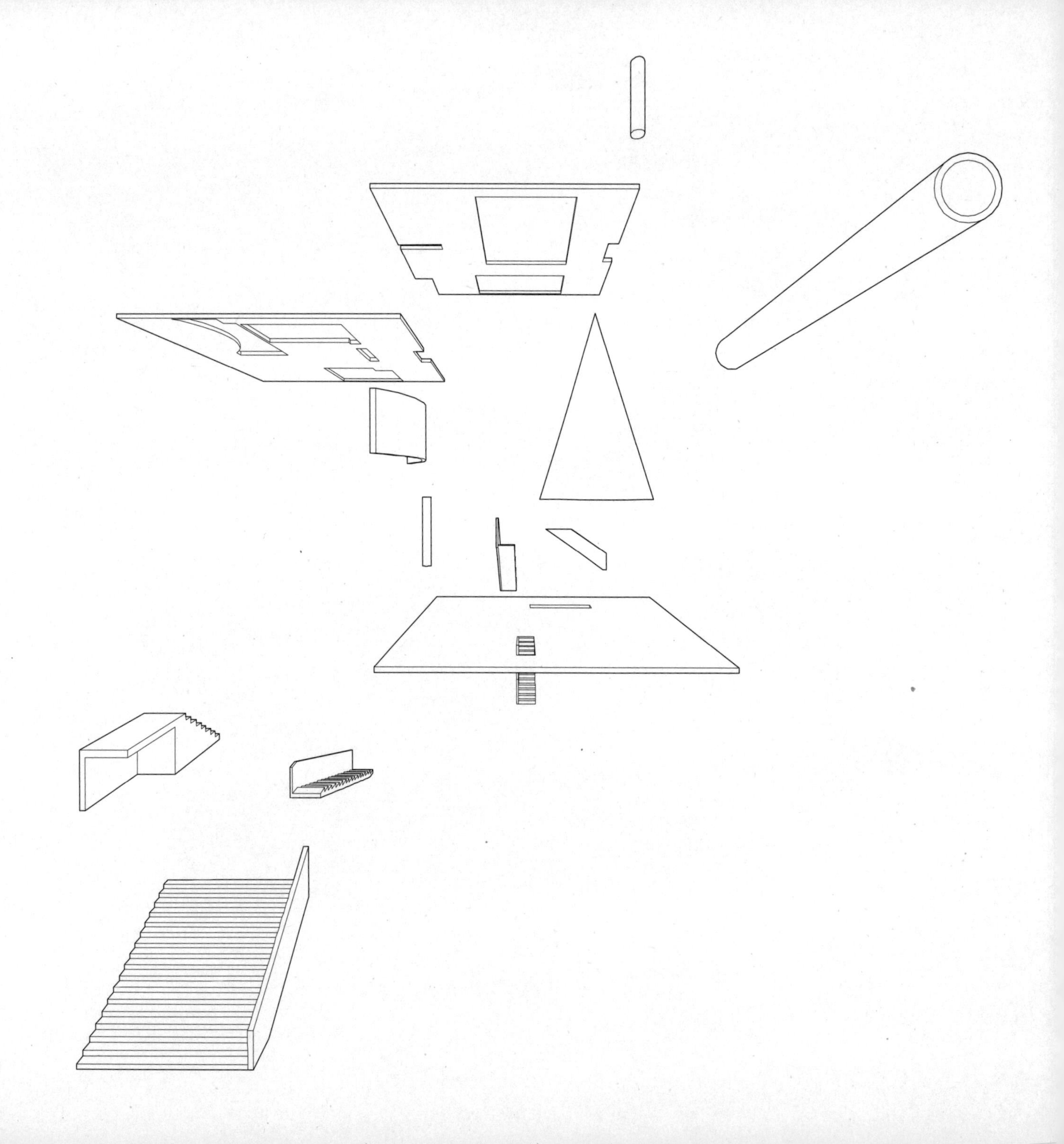

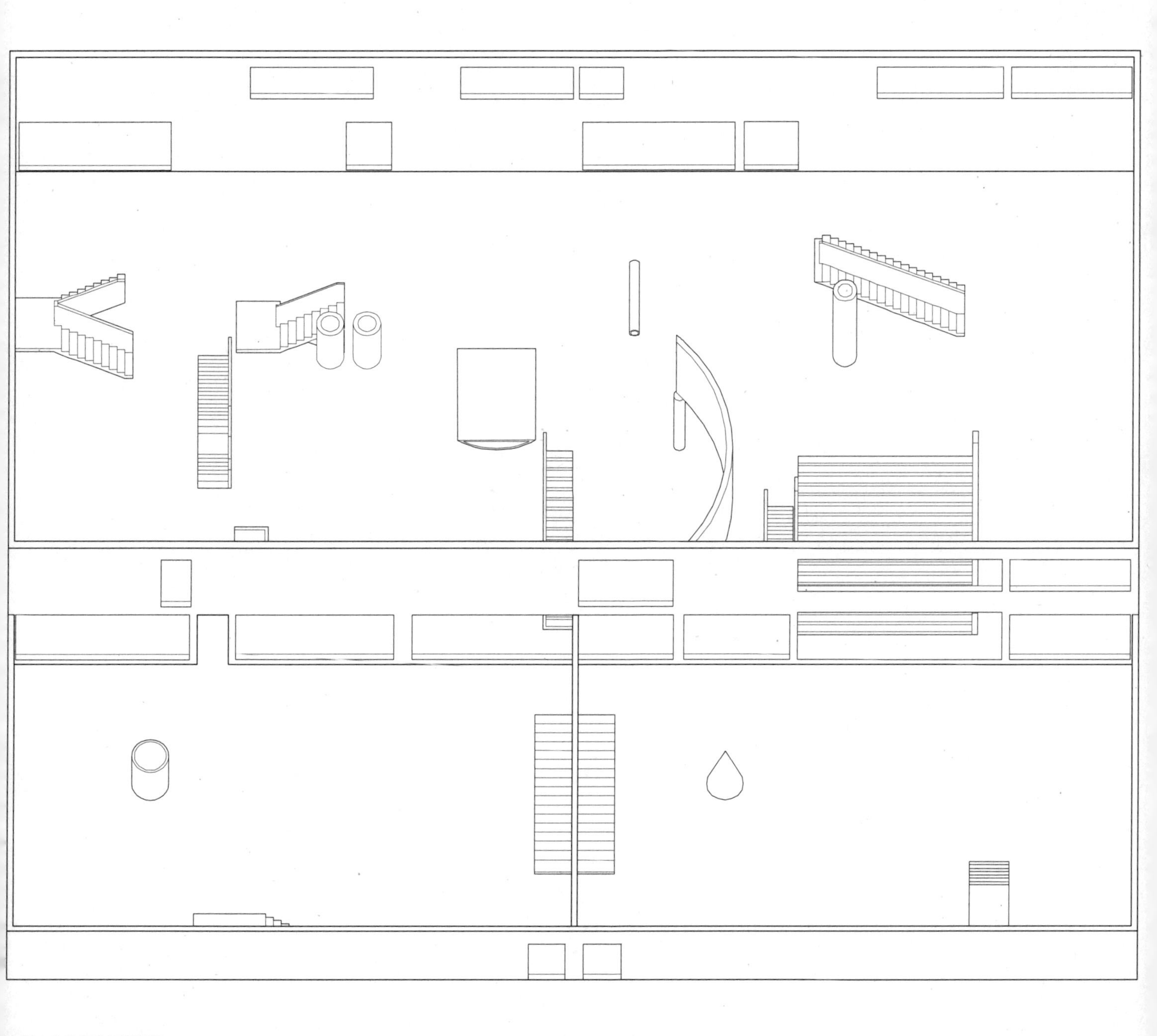

左图：住宅内的几何要素
上图：箱体空间内的要素布局
Left Figure：geometric elements in the house
Up  Figure：layout of the elements in the space-box

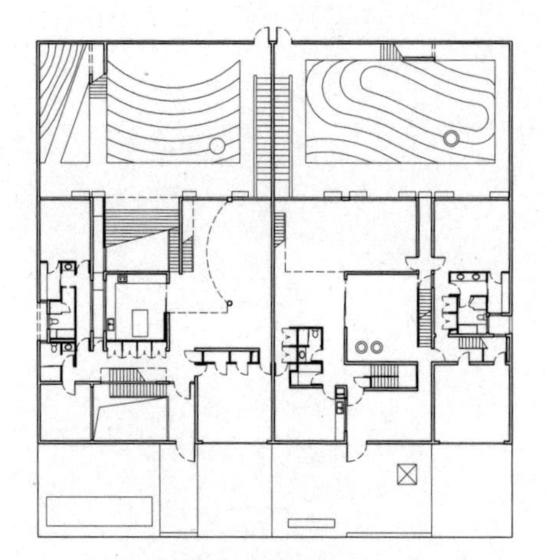

首层平面图 / The 1st floor plan

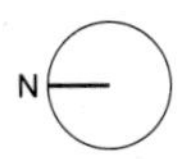


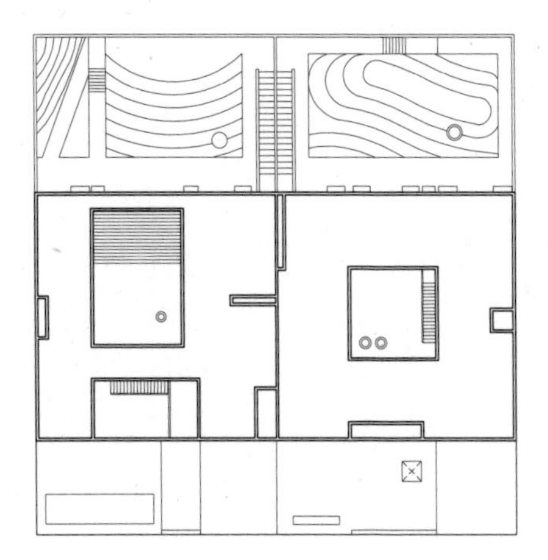

屋顶平面图 / Roof floor plan

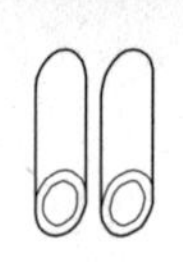

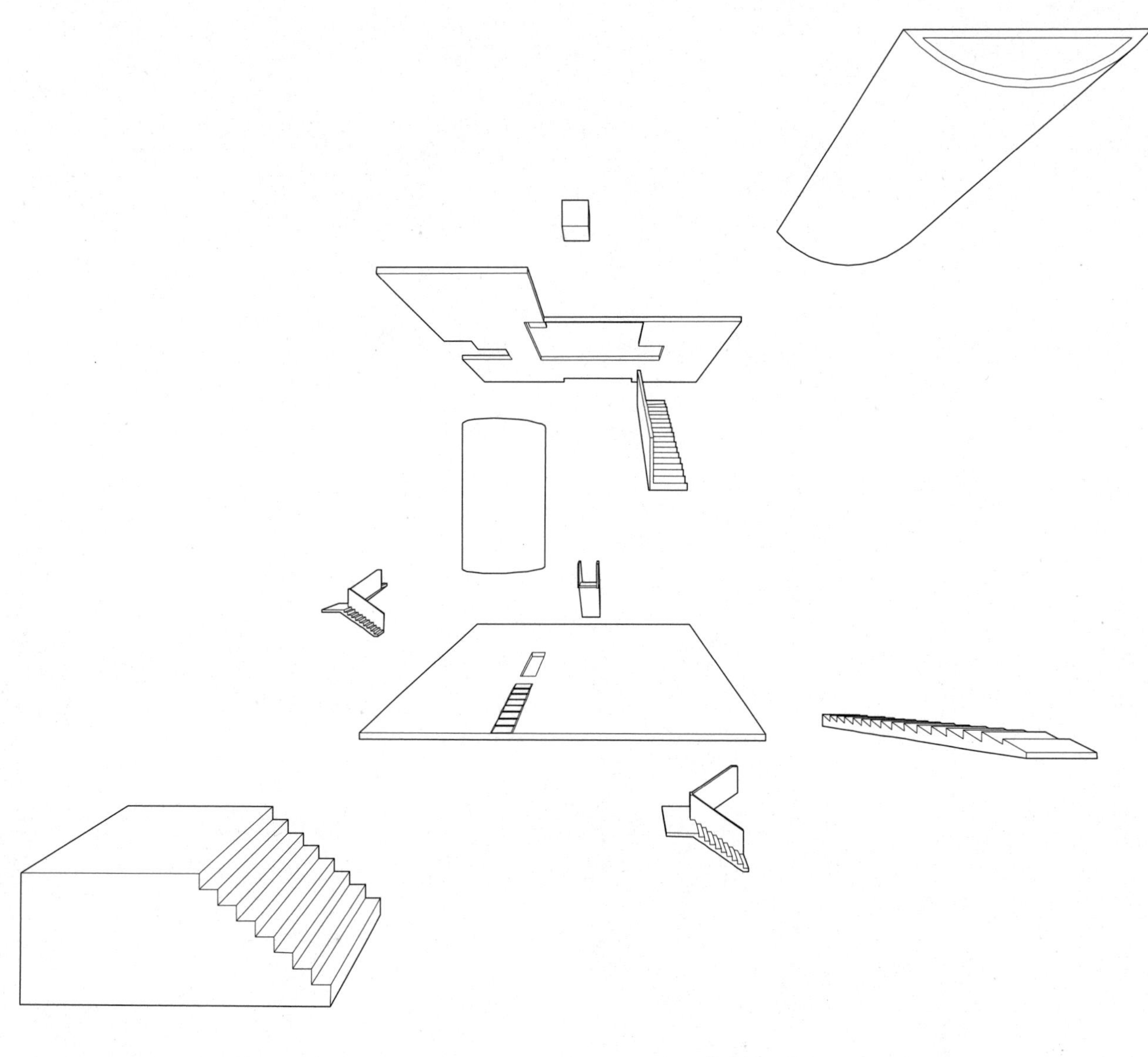

王昀
Wang Yun

60平米极小城市

$60m^2$ Mini City

Wang Yun

王昀

2003

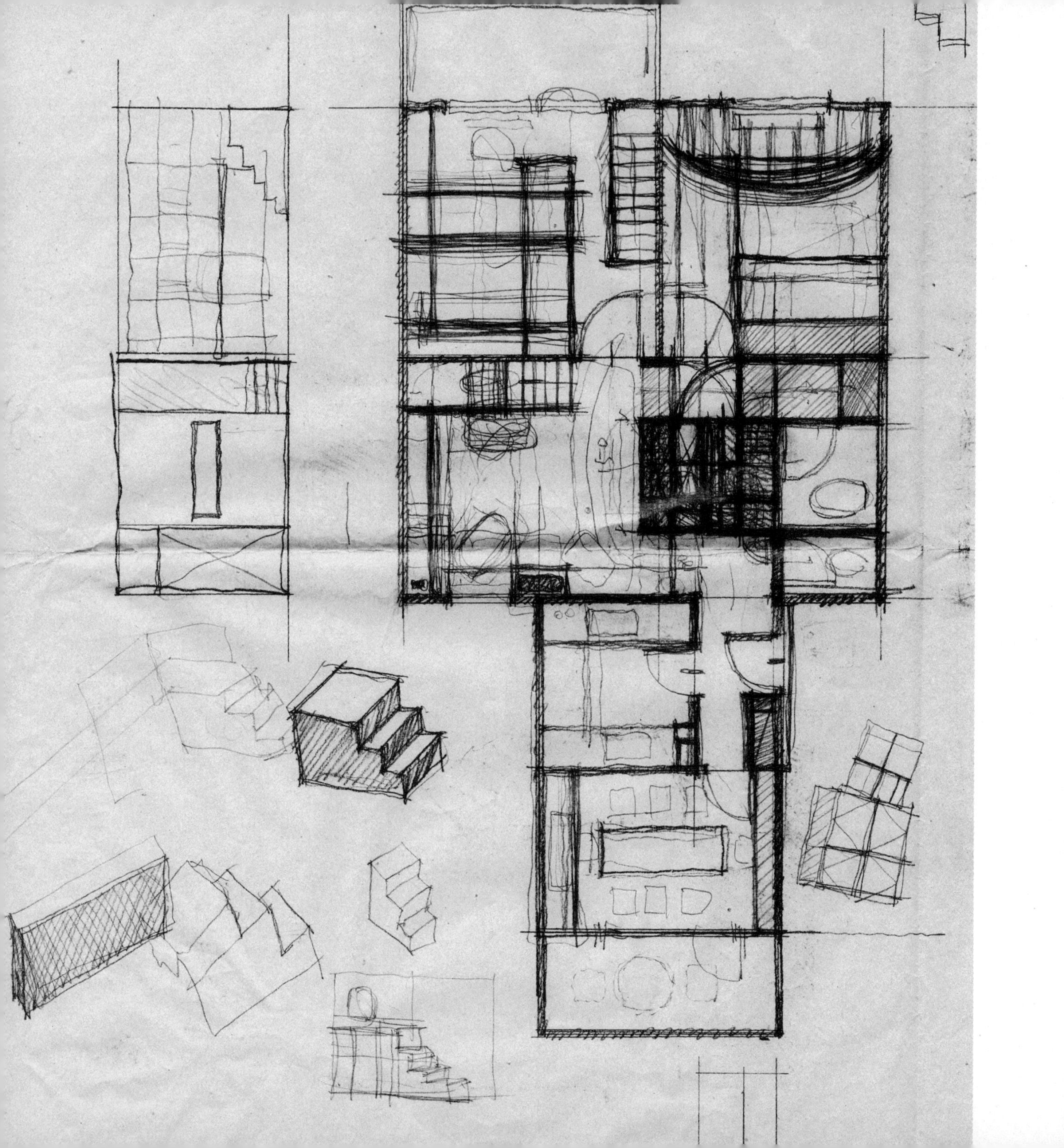

60平米极小城市的构思草图（2001年）
60m$^2$ Mini City Concept Sketch (2001)

# 1

## 60平米极小城市
60m$^2$ Mini City

2001年从北京大学租借来的三室一厅住宅位于整体为6层的砖混结构建筑的三层。此住宅楼为一梯两户，南侧有两间居室，北侧有一间。住宅入口的对面是厨房，厨房的南侧是客厅。在没有改造之前，从住宅外部进入内部在空间上没有任何的“渐变”与“过渡”。其中，住宅中卫生间的门被设计成直面客厅的处理是最不能忍受的。这个住宅的可使用部分只有60平方米。由于墙体均为承重结构，所以无法进行根本性的调整。因此在进行新的改造时，力图在自己未来居住的内部空间中将符合人心理空间渐变的尺度关系加以还原确定为设计的中心思考问题。尽管房子的物理空间很小，但实际上其构造关系与城市是相互对应的。如起居室与城市广场对应、餐厅与城市餐厅对应、卧室与城市的旅馆对应、书房则与图书馆等公共场所相对应。而将所有这些功能之间加以串并联是街道的功能所在。考虑到这些场景，当我们将S（极小）尺度的室内家具转换为城市的XL（极大）尺度时，一个与城市构造相对应的60平米极小城市的风景便由此而展开。

It's a 3-bedroom flat on the 3rd floor of a 6-floor brick-concrete building, which I rent from Peking University. Two flats on each floor. Two bedrooms in the south, one in the north, kitchen opposite to entrance with living room to its south. There wasn't any gradual change or transition from outside to inside. Major problem is bathroom opens to living room. The usable floor area is only 60m$^2$ and all walls are bearing structures so essential adjustment is impossible. Therefore how to make my residence scale fit the psychological space transition is the renovation's top priority. Although the physical space is limited, its structural relation corresponds to the city, such as living room matches urban plaza, dining room matches urban restaurant, bedroom matches urban hotel, and study matches library. And it's the streets' role to connect all these functions. All scenes considered, when we change the interior furniture's S size to the city's XL size, a 60m$^2$ Mini City appears.

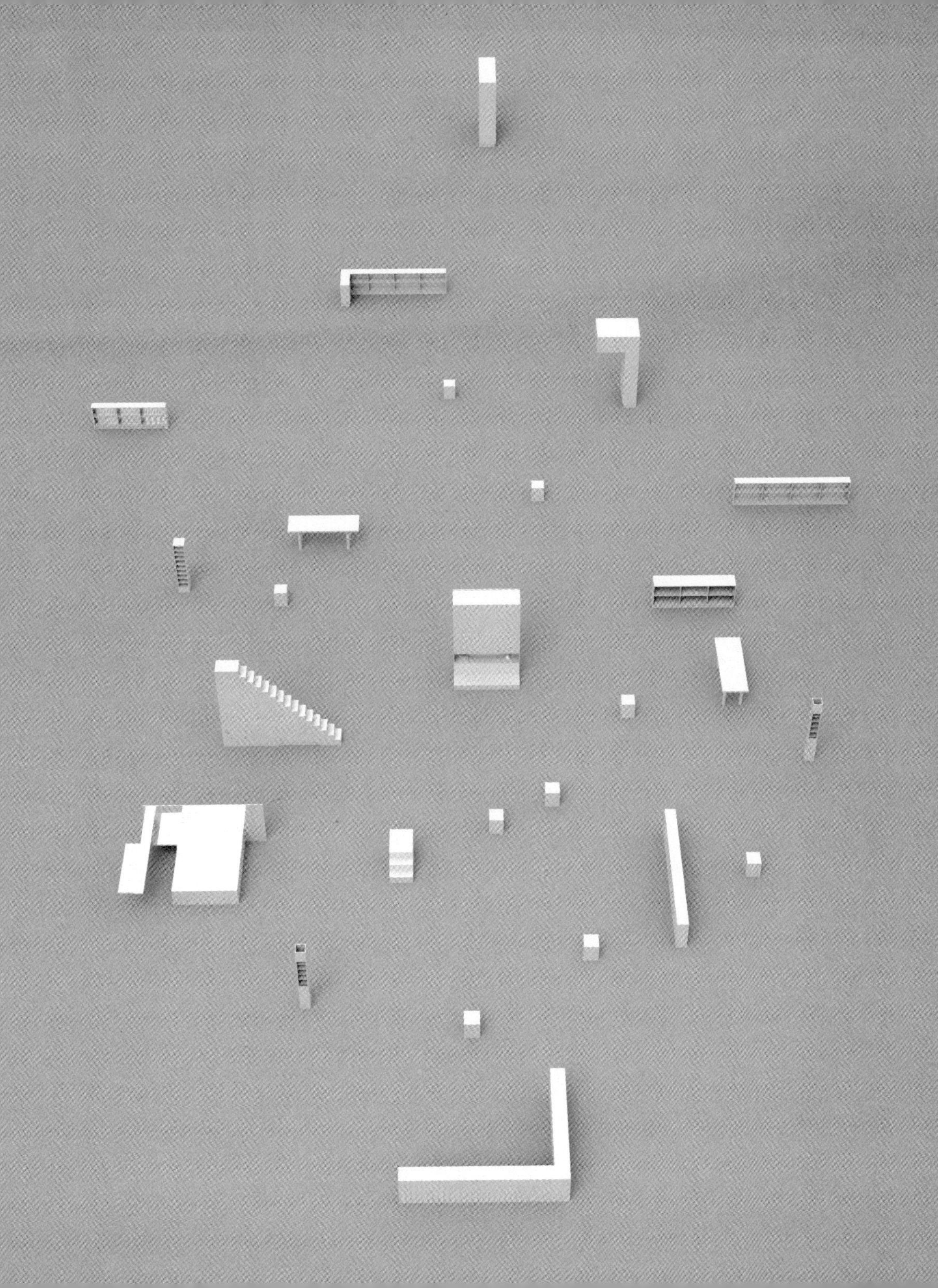

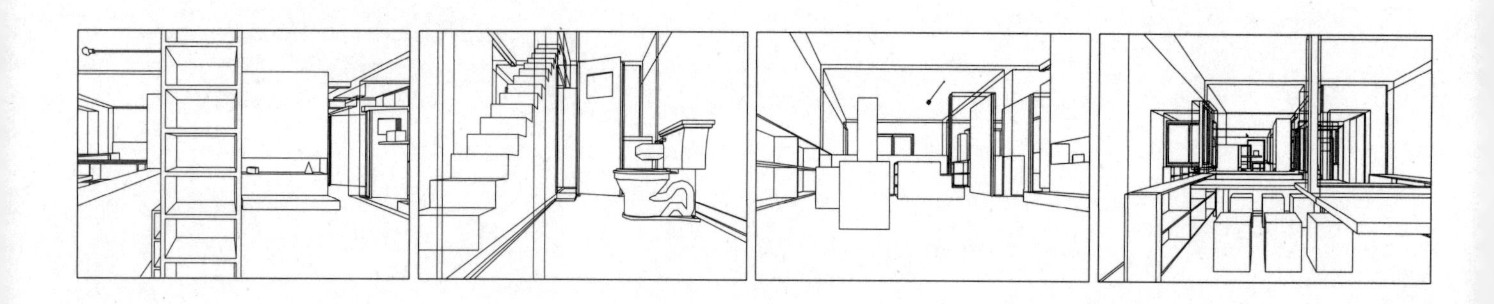

记忆中的桃坪村风景
Taoping Village's scenery in memory

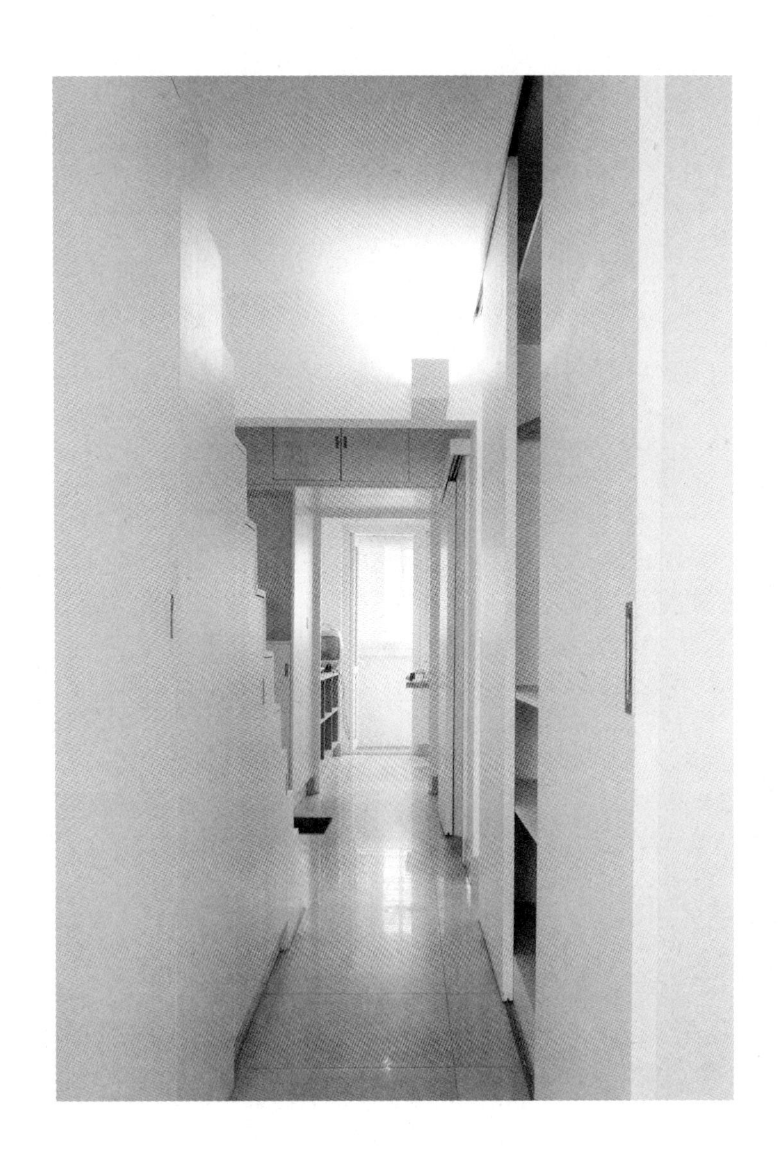

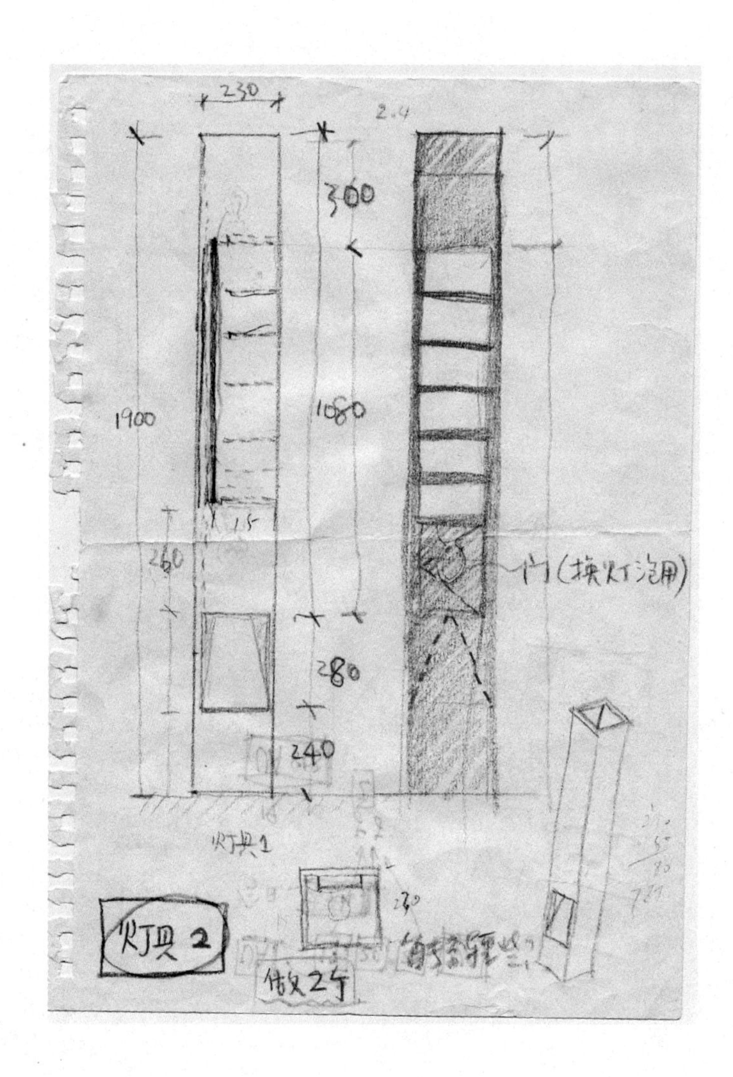
230
300
1900
1080
260
280
240
门(换灯泡用)
灯具1
灯具2
做2个

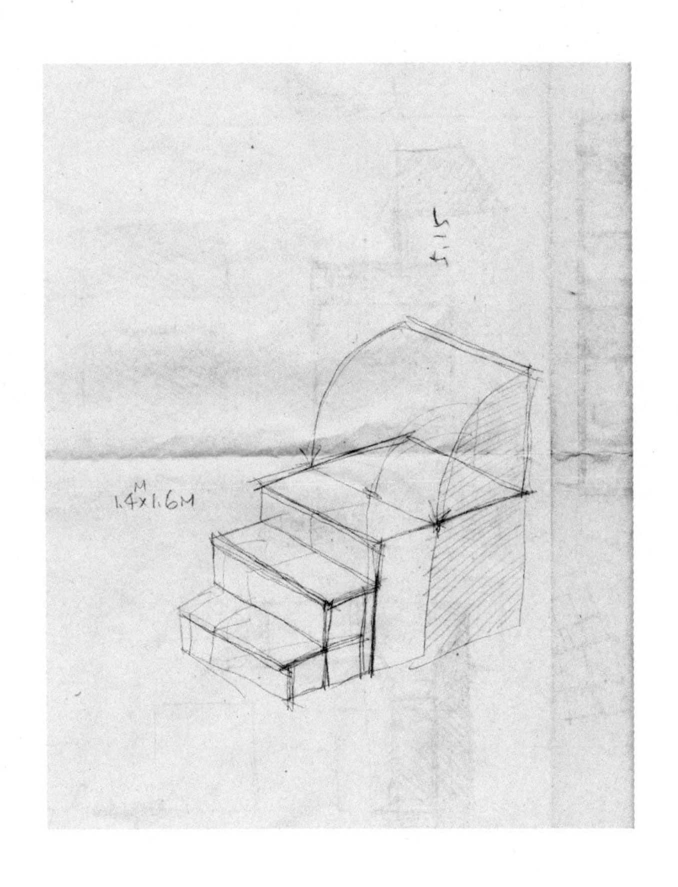
1.4$^{M}$x1.6M

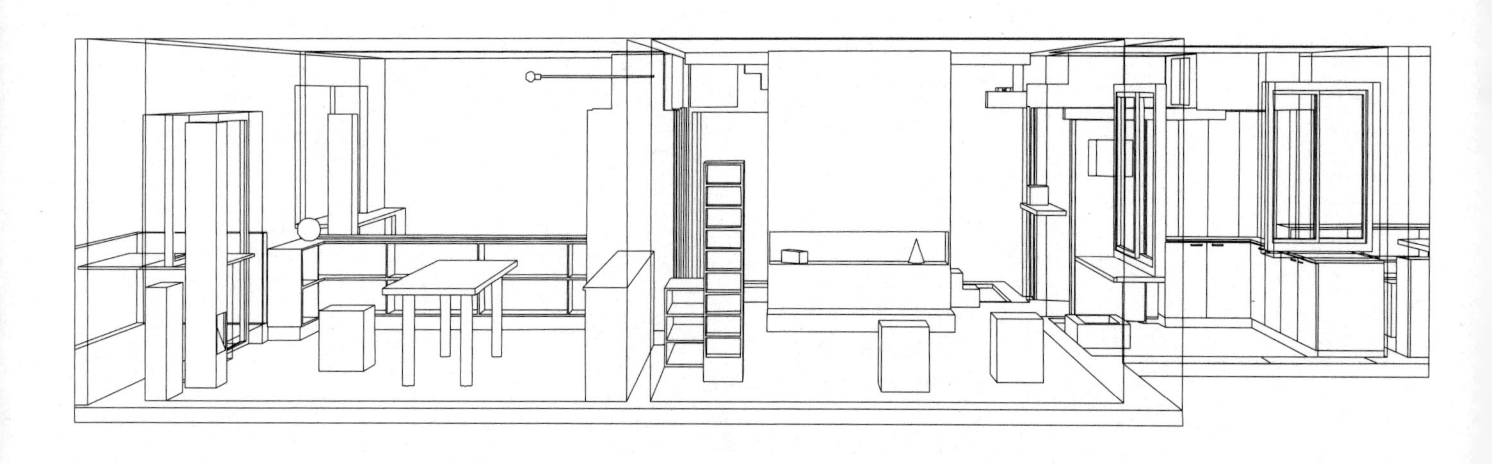

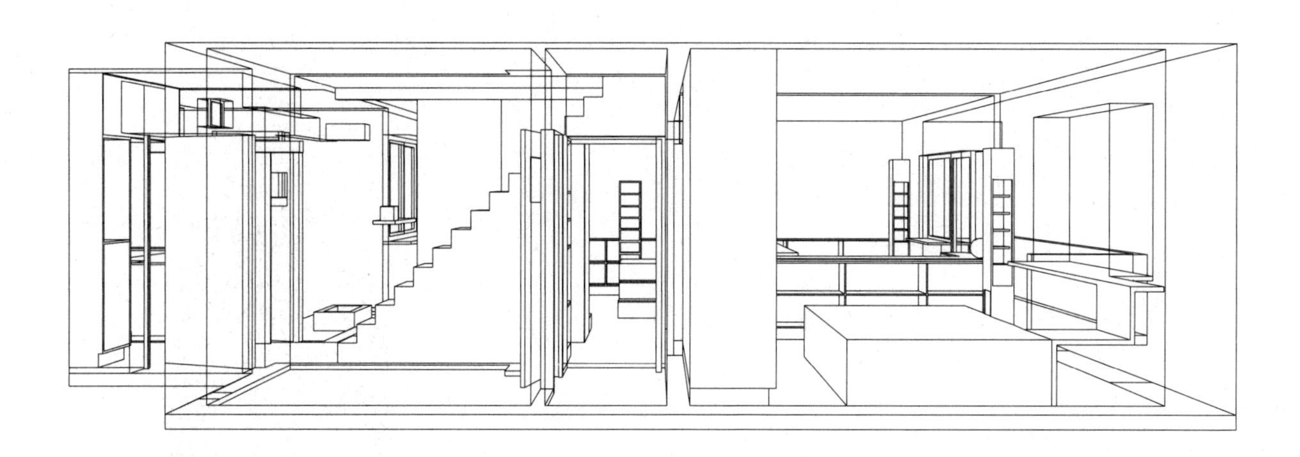

# 2002

王昀

# 门与厅

WANG YUN

# Door and Hall

从摩洛哥的马拉喀什通往塔伦丹特(Taroudant)路边的住居，白色涂料泼洒式地喷涂在建筑的关键位置，构成了门与厅的意向风景
Houses alongside the road from Morocco Marrakech to Taroudant. White paint splashes on the major part of buildings, which constitutes the view concept of Door and Hall

# 门与厅

Door and Hall

门与厅是一个在原有办公楼东侧进行增建的小项目。基地南北两侧有两栋建造于不同时期的住宅楼，南侧住宅楼为红砖色，北侧住宅楼为米黄色，原有场地内还有几个零散的小仓库，一个狭长的窄道使基地与城市道路相连，场地整体色调斑斓，气氛凌乱。为了强调入口的主体特征，我们将原有米黄色的办公楼施以黑色，使其淡化并与周围的住宅楼共同成为围合办公楼前广场空间的背景要素。增建的门厅采用单纯的白色几何形体，运用摆放一个微弯曲方筒体块的方法使之成为小广场上的视觉中心。

It is a small extension project to the east of an office building. A brown apartment building in the south of the site,and a beige apartment building in the north, with a few warehouses scatter around. A long narrow alley connects the site to the urban road. The whole site's colors and atmosphere are in disorder. To stress the entrance's characteristic, we painted the original beige office building into black to blend it into plaza background together with surrounding apartment buildings. The newly built foyer uses pure white geometric objects and set a slightly curved cubicle as the little plaza's visual center.

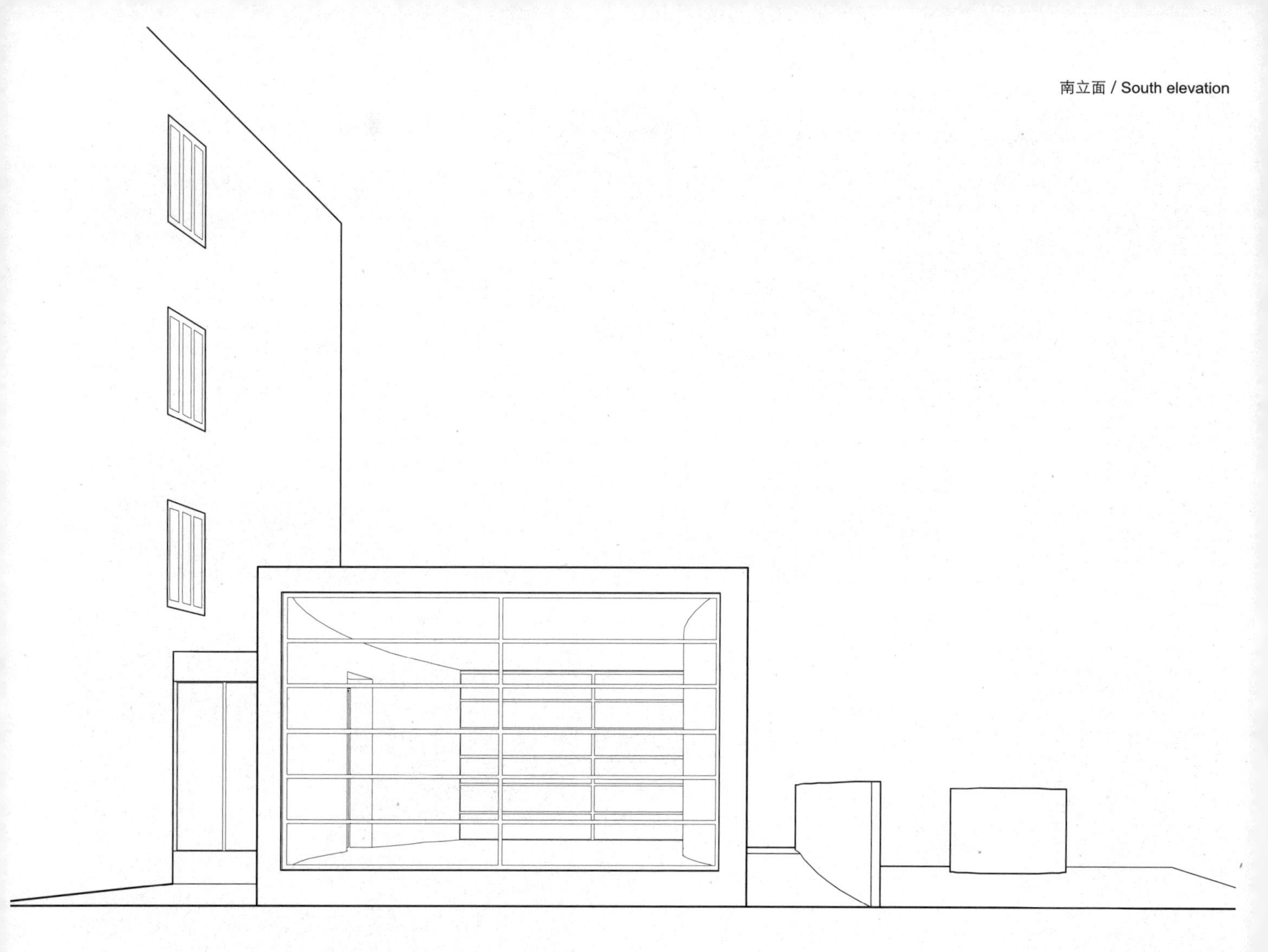

南立面 / South elevation

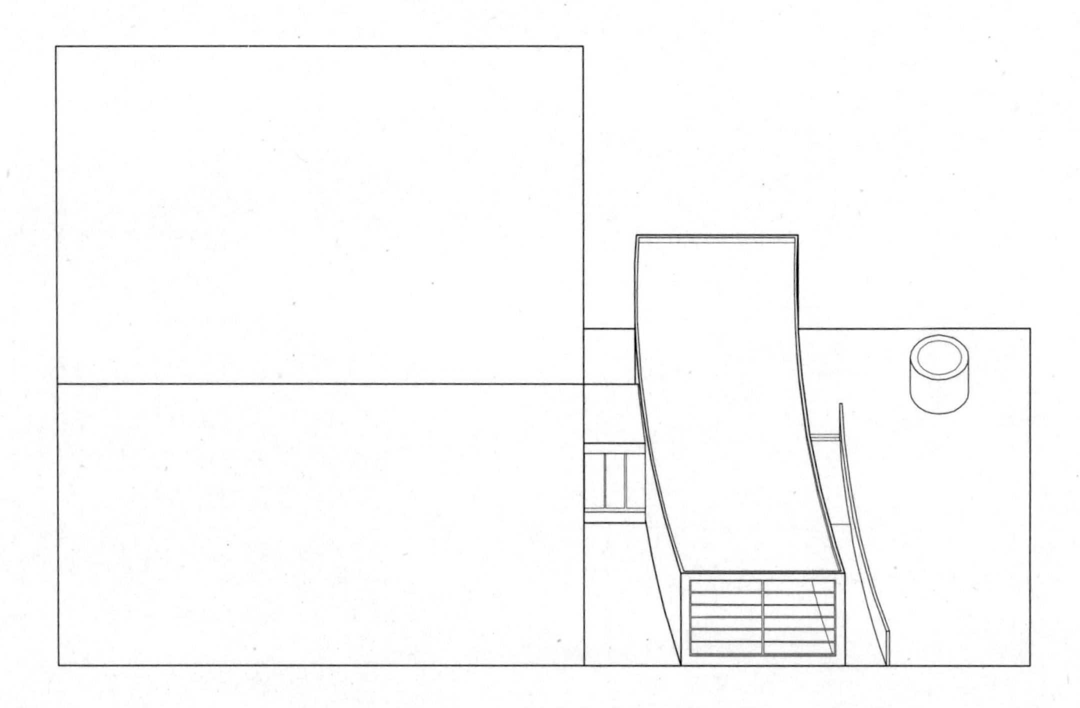

轴测图 / Axonometric drawing

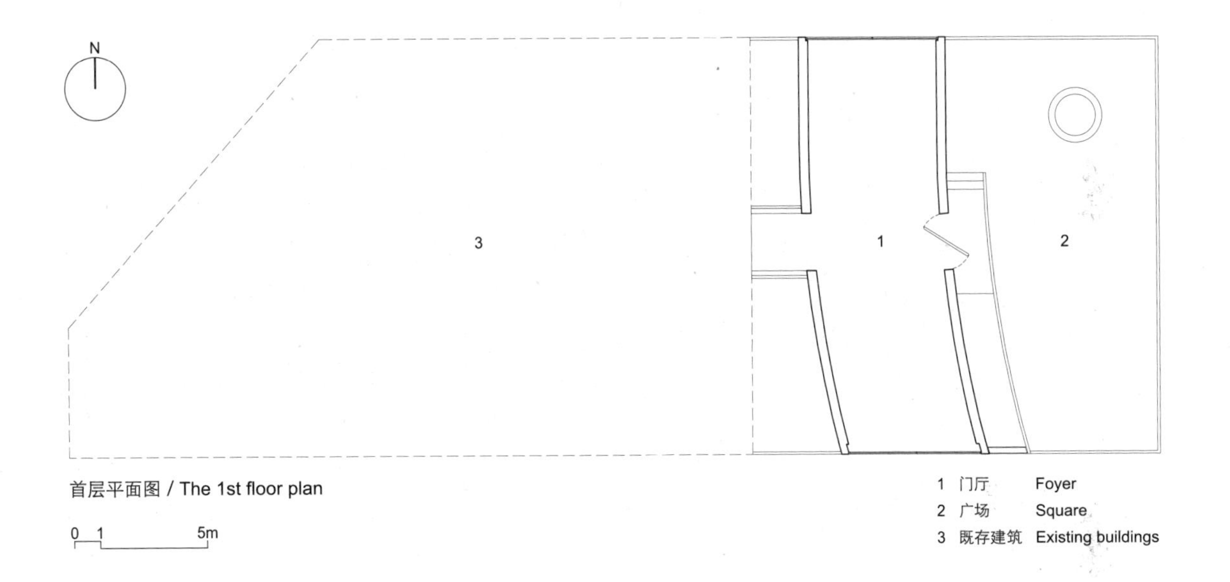


首层平面图 / The 1st floor plan

1 门厅 Foyer
2 广场 Square
3 既存建筑 Existing buildings

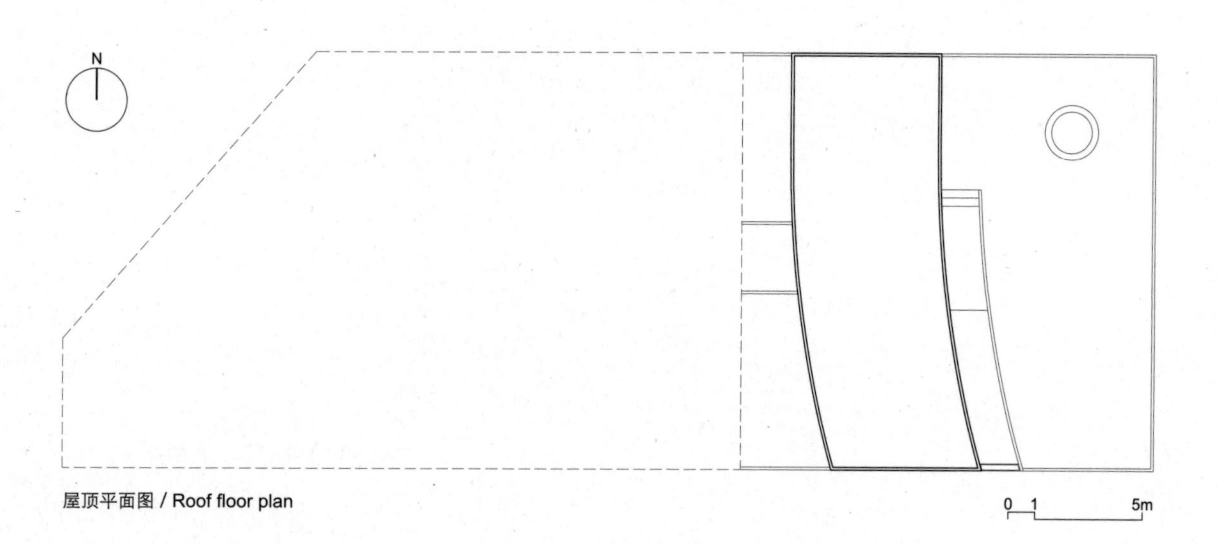

屋顶平面图 / Roof floor plan

拉

# 财政局
# Finance Bureau

王昀

wangyun

2004

## 财政局
## Finance Bureau

石景山财政局建筑面积为7000平方米。考虑到未来办公人员的舒适度，设计时将所有办公室门与走廊面向一个巨大的中庭，目的是使得每一个工作人员在办公室的出入之间获得收与放的心情调节和气氛转换。建筑被设计成宽与高均为40米，长为60米的体块。位于中庭上部及其西侧墙面上的46个3米直径的圆形采光孔，在为中庭内部倾泻阳光的同时还突出了“凿”的概念。

Shijingshan Finance Bureau's floor area is 7000m$^2$. Taking future staffs' comfort into consideration, all the office doors and corridors are designed to face a giant atrium, to enable mood adjustment and atmosphere change when staffs go in and out their offices. 40mx40mx60m block with 46 3m-diameter round lighting holes on atrium ceiling and west wall, provide sufficient sunlight for the atrium. In the meantime, the concept of "chisel" stands out.

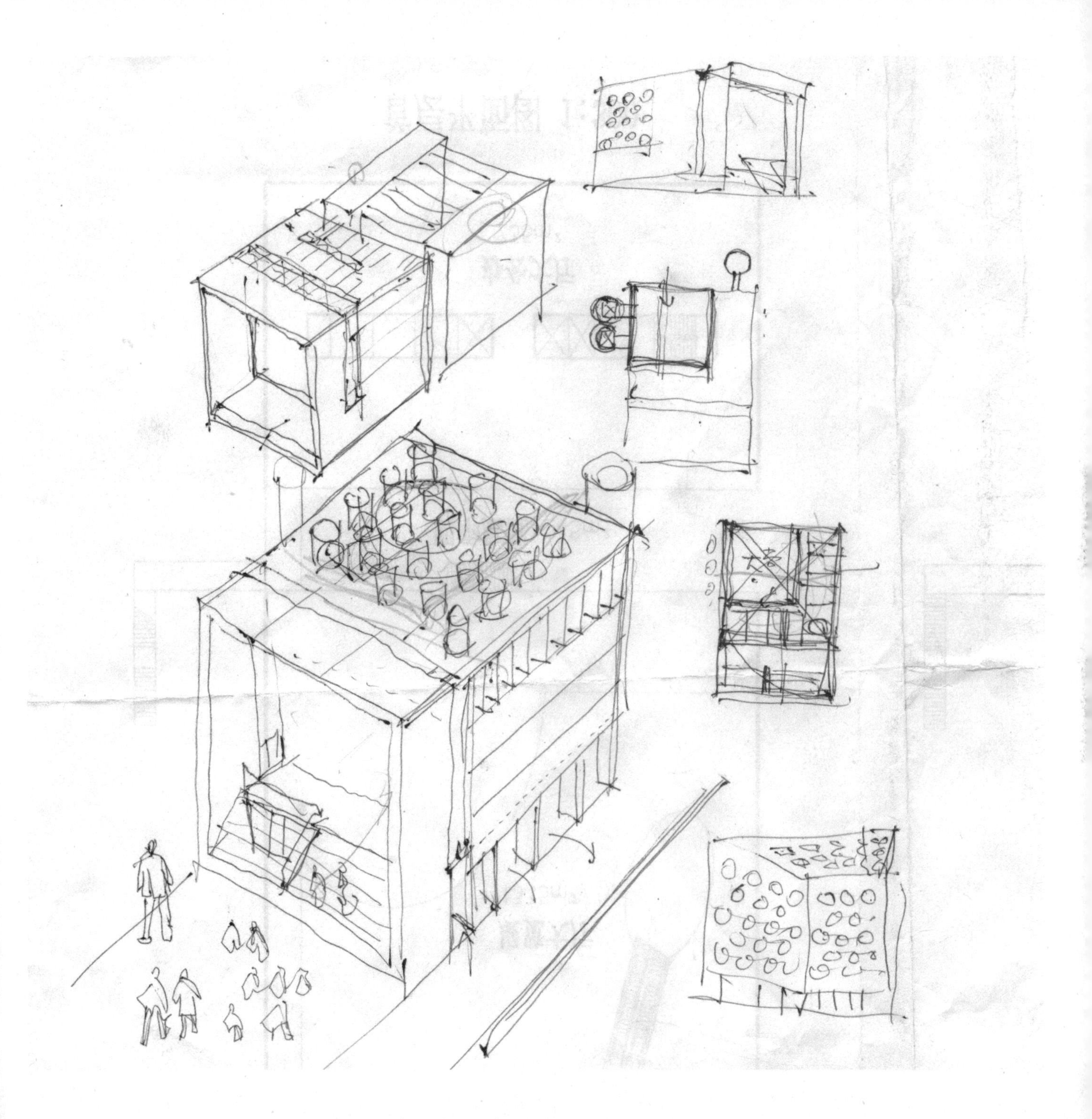

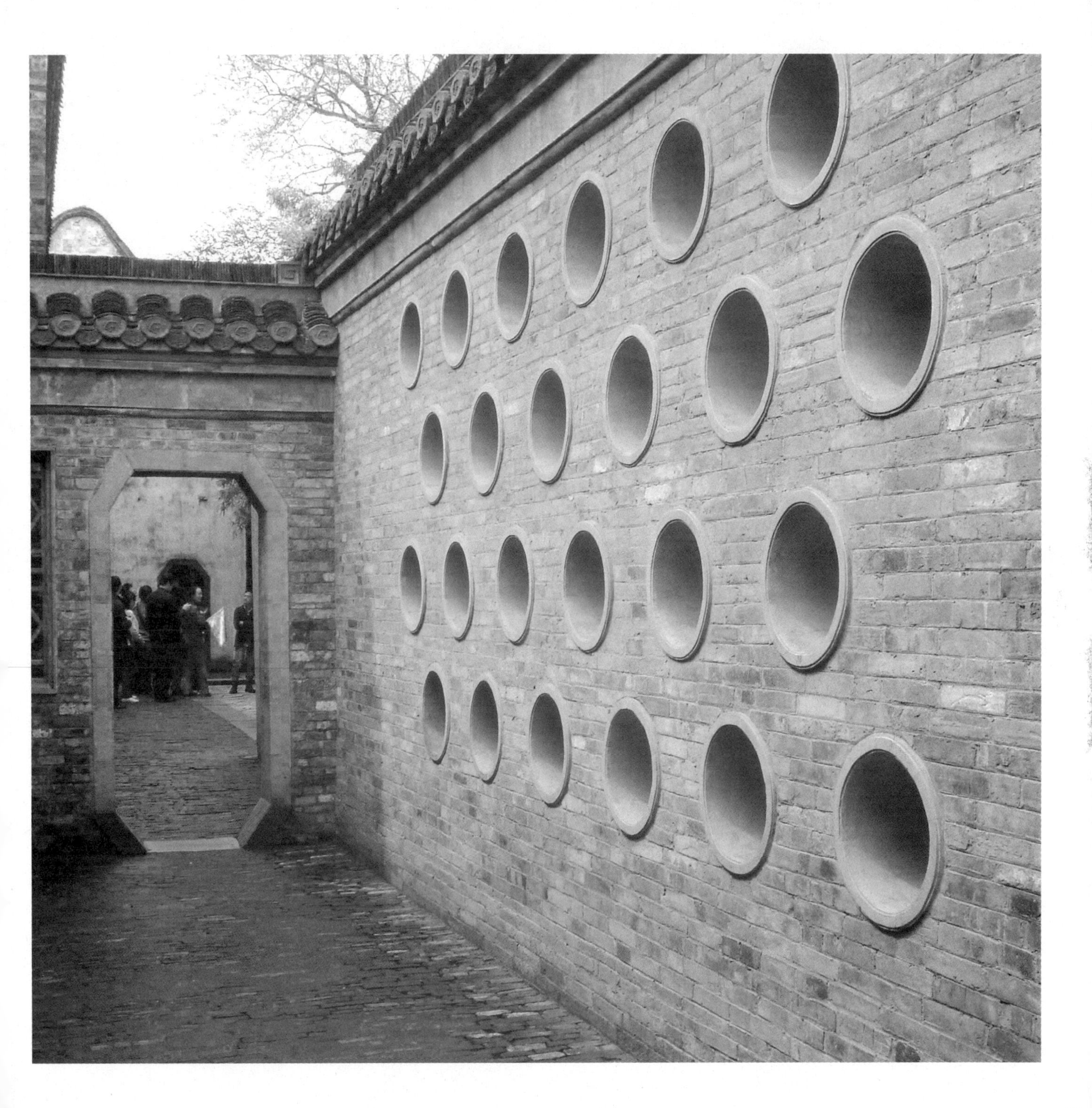

轴测图/Axonometric drawing

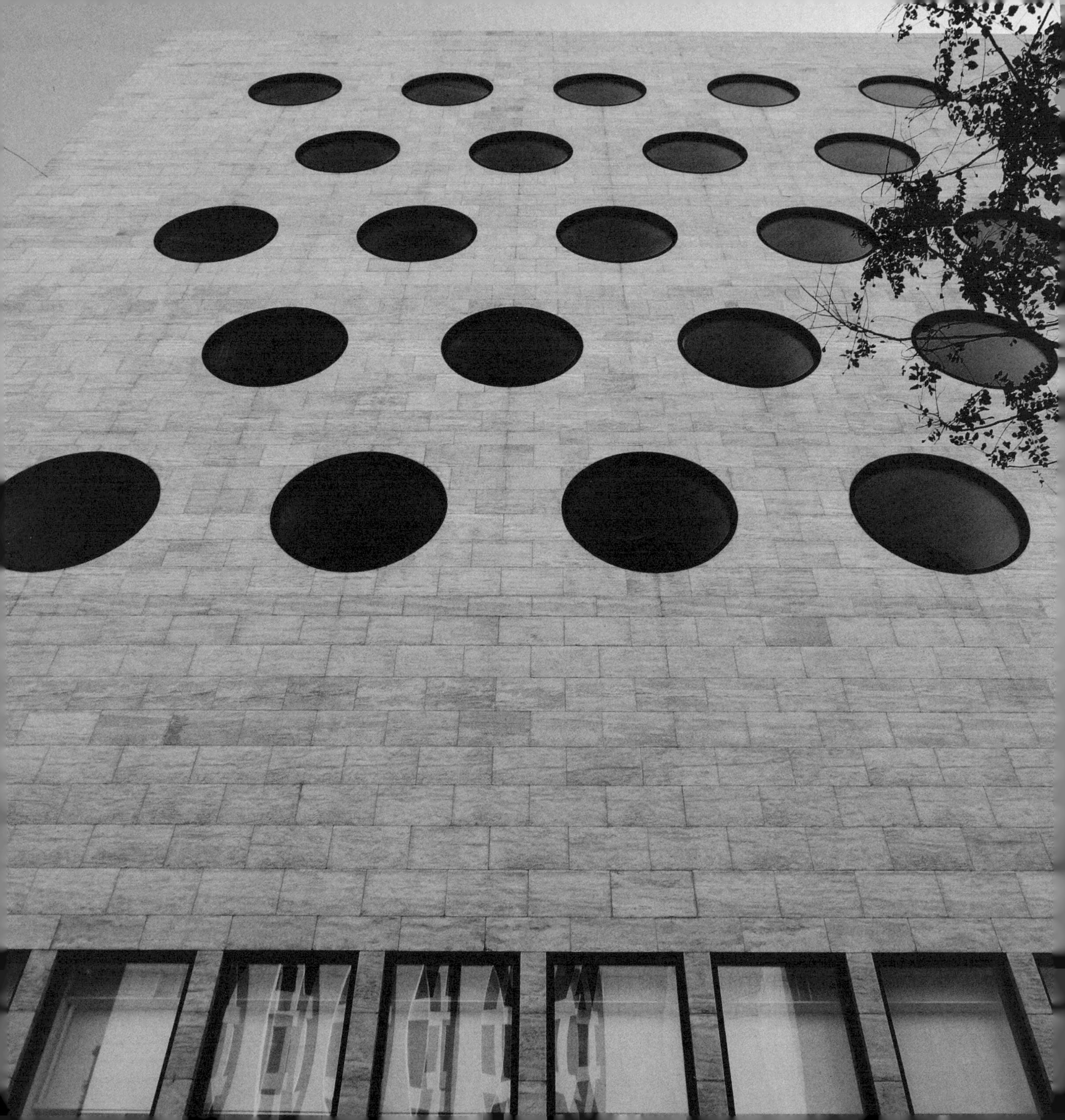

首层平面图 / The 1st floor plan

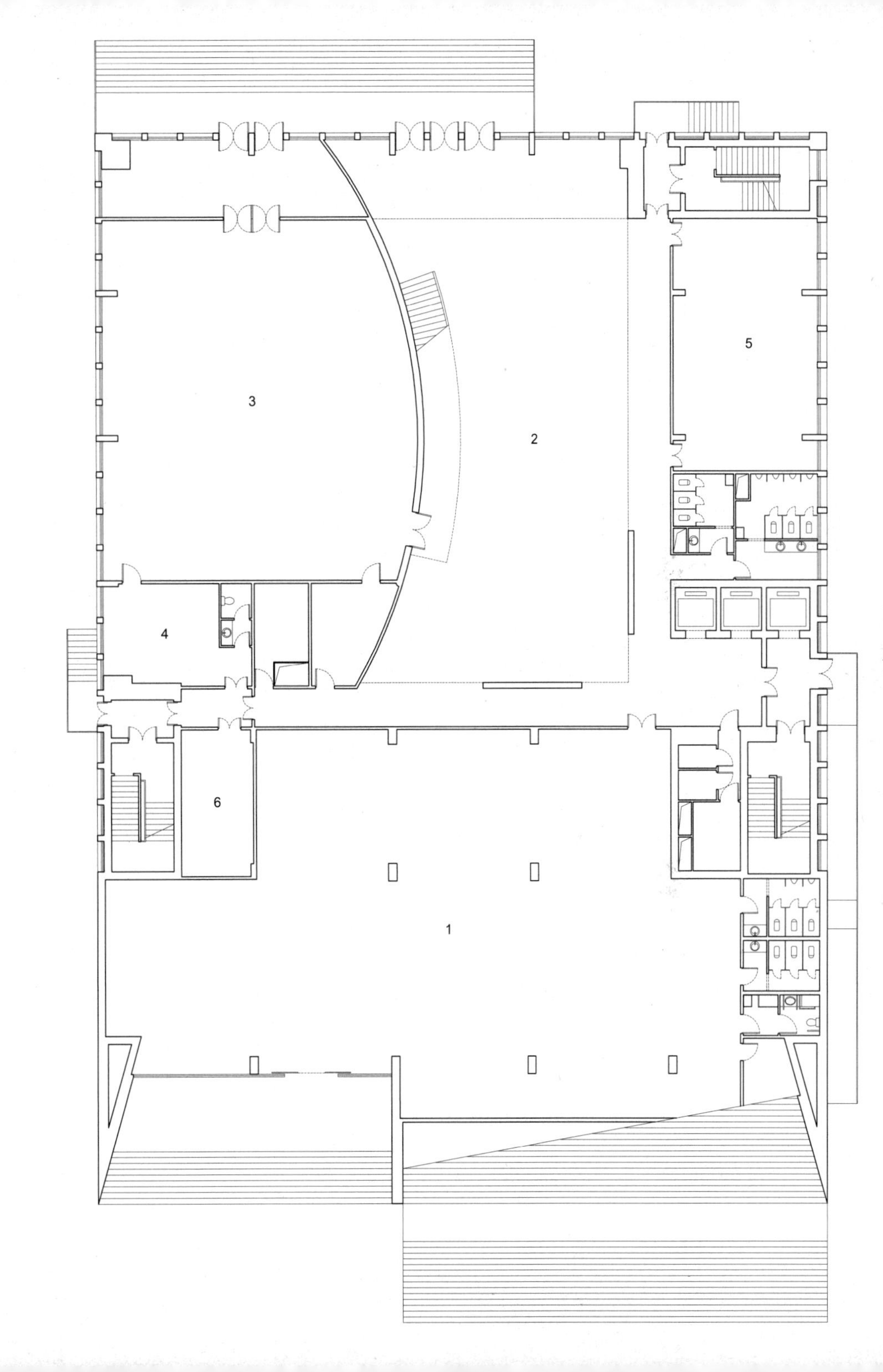


1 营业大厅 Business hall
2 大堂 Hall
3 多功能厅 Multi-function hall
4 贵宾厅 VIP hall
5 培训教室 Training classroom
6 休息室 Rest room

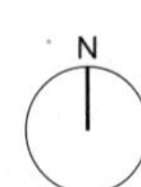


0 1 5m

二层平面图 / The 2nd floor plan

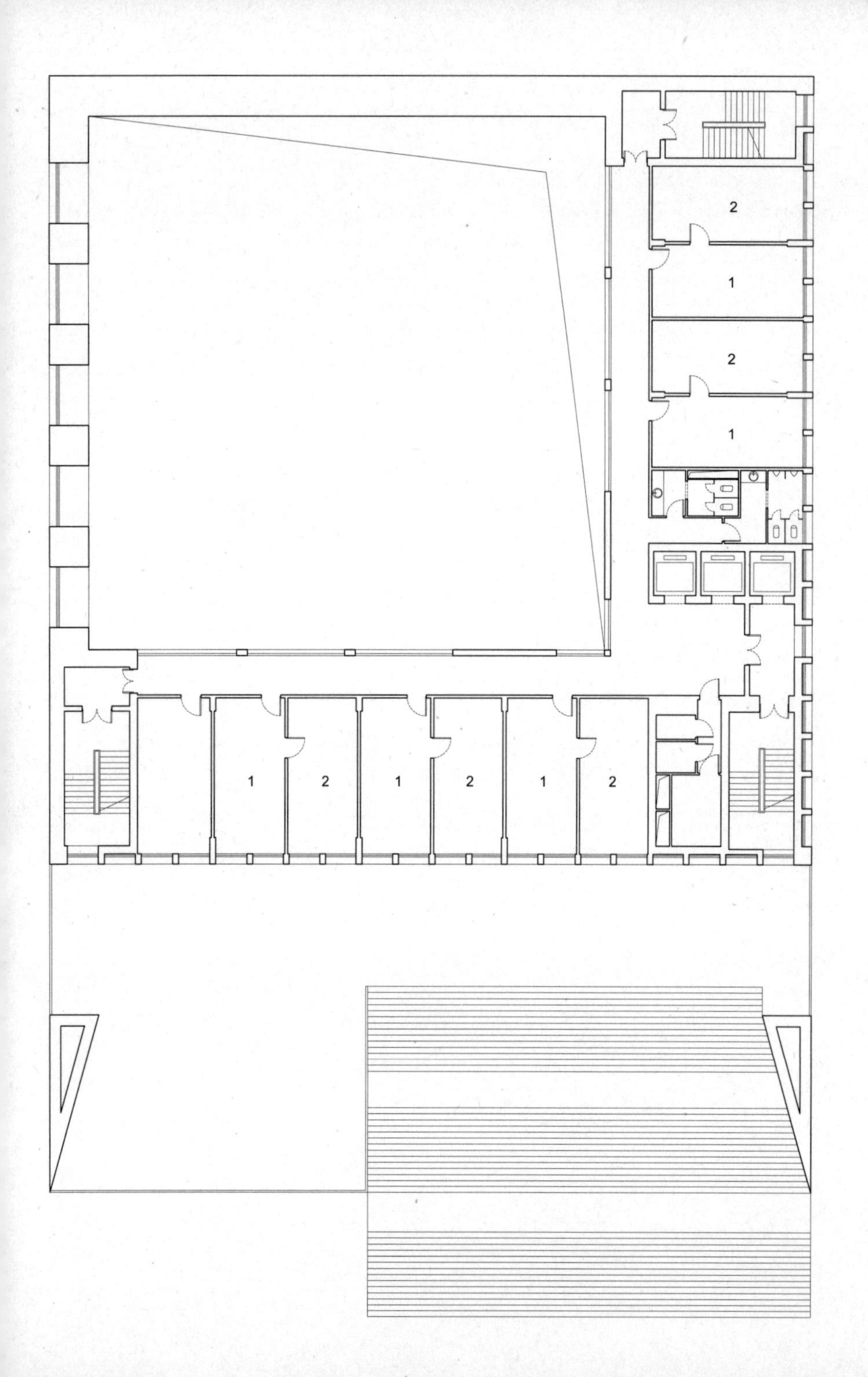


1 科员室 Staff room

2 科长室 Section director room

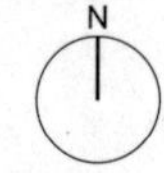


0 1 5m

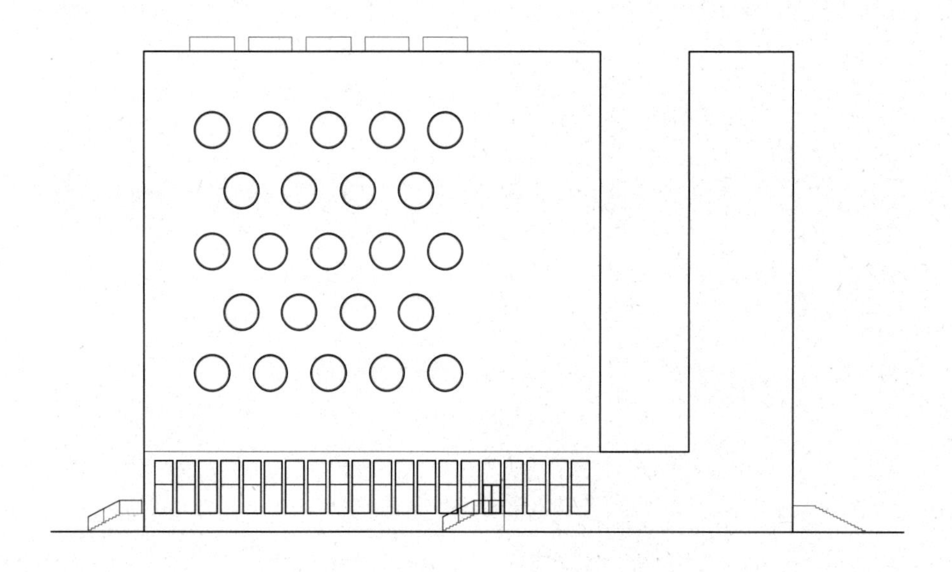

西立面 / West elevation

“凿”的概念本身还是中国传统哲学思想的体现。中国古代哲人老子所谓“凿户牖以为室，当其无有室之用”的论述，使该建筑本身在“凿”的概念上获得地域性意义的解读。

"Chisel" embodies traditional Chinese philosophy. A famous quote by Lao Tzu: "We chisel doors and windows to make a house; but it is the inner space that makes it liveable", gains a regional illustration here.

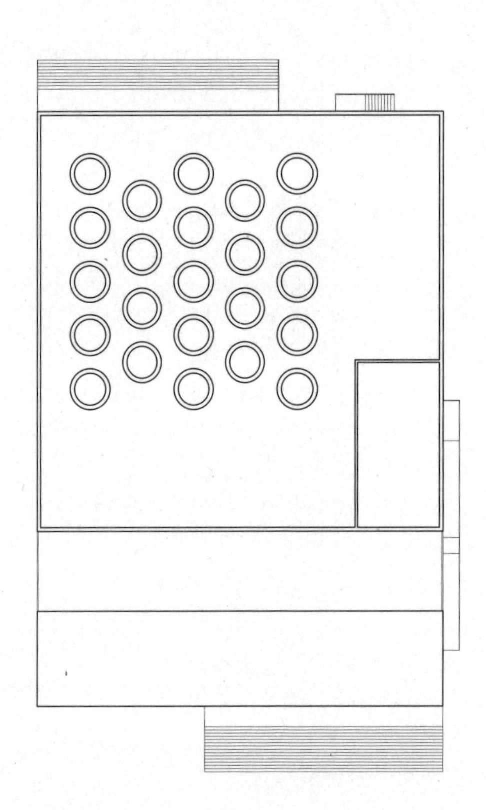

# 会所
# Club House

王昀

WANG YUN

2004

摩洛哥提·兹乌林（Tin Zoulin）聚落住居是一个围合式的巨大院落，令人联想起中国福建的方楼，同时也易联想起山西与河南的地下窑洞以及北京四合院。

Morocco Tin Zoulin settlement is a huge enclosed courtyard, reminds people of Fujian Tulou, Henan underground cave dwelling and Beijing courtyard.

# 会所 Clubhouse

庐师山庄会所是为52栋别墅园区提供的一个公共活动场所，建筑面积为1600平方米。建筑共3层，内部设置有咖啡厅、会议室和4间客房。会所的入口是一个三层高的大厅，大厅与室外以整面的玻璃幕墙相隔，室内与室外之间互为对象的对景与互位关系是该建筑设计的主旨。会所西向面对的是整个区域的中心绿地，这个绿地保存了基地上原有的园林树木，具有良好的景观。为此，沿着面向中心绿地景观的展开面方向，设计时在大厅设置一个直接通向二层的坡道，这个坡道能让人们在逐渐上升的过程中体验到西侧大玻璃幕墙之外逐渐升起的景象。会所西侧的整体采用吊挂系统的高11米的幕墙，由于幕墙上部分的玻璃7米高，现实中的钢化炉只能提供6.8米的钢化玻璃，使这个幕墙无意中挑战了时代的极限。

Lushishanzhuang Clubhouse is the public building for the 52 villas within the compound. The three-floor clubhouse has a total floor area of 1600m$^2$. There are cafe, meeting room and 4 guest rooms. The entrance is a three-floor-tall hall with a one-piece glass curtain wall. The design intends to make indoor and outdoor complement one another through the transparent glass. The clubhouse's west facade faces the central green space of the whole area. The original trees are preserved so the landscape is very pleasing. Therefore, alongside the green space, I designed a sightseeing ramp leading to the second floor. The escalating passage will enable people to experience the changing view outside. The west overhang curtain wall is 11 meters tall class. The upper part of class is 7 meters tall while today's toughening furnace can only produce 6.8 meters tall, making this curtain wall by chance challenge the era's technique limit.

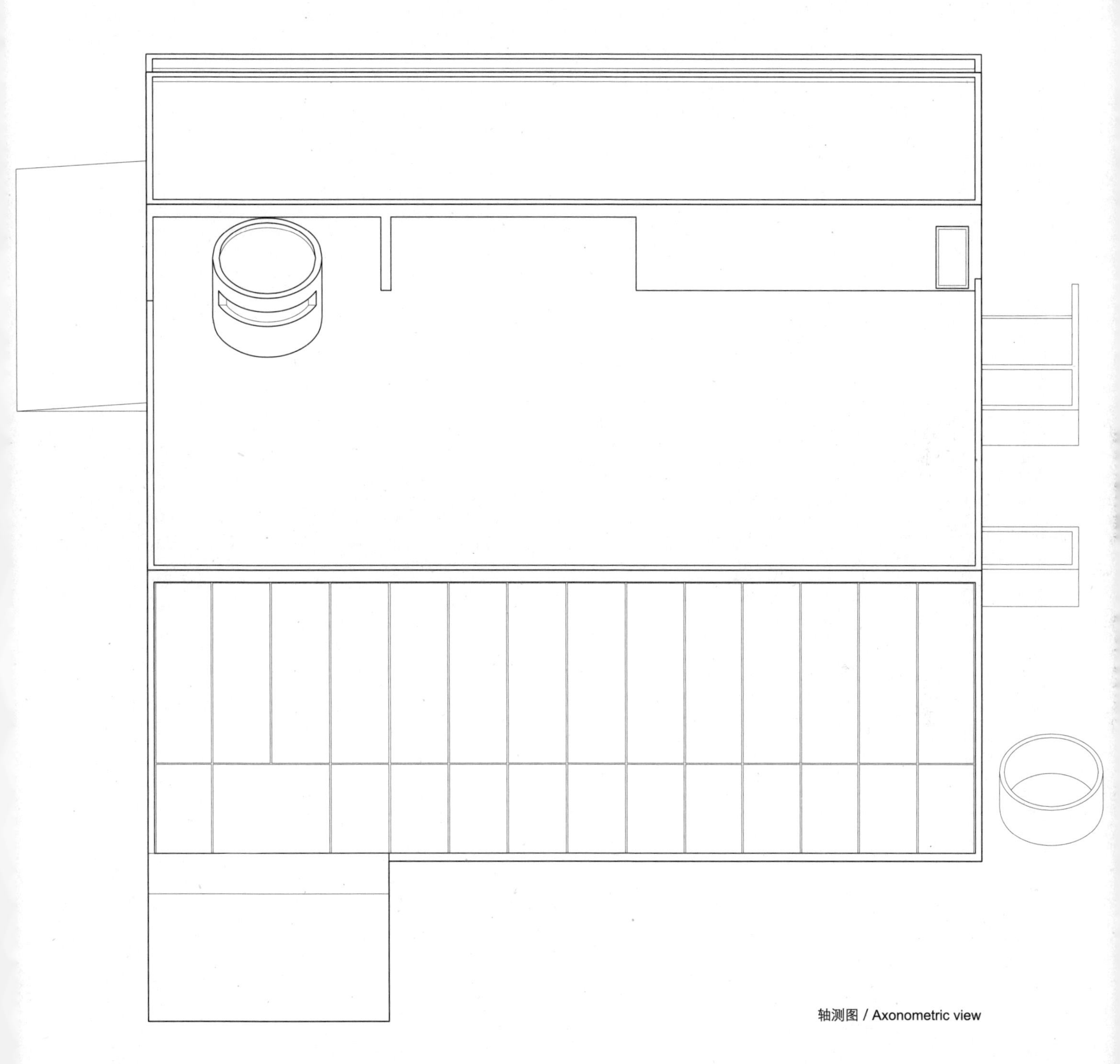

轴测图 / Axonometric view

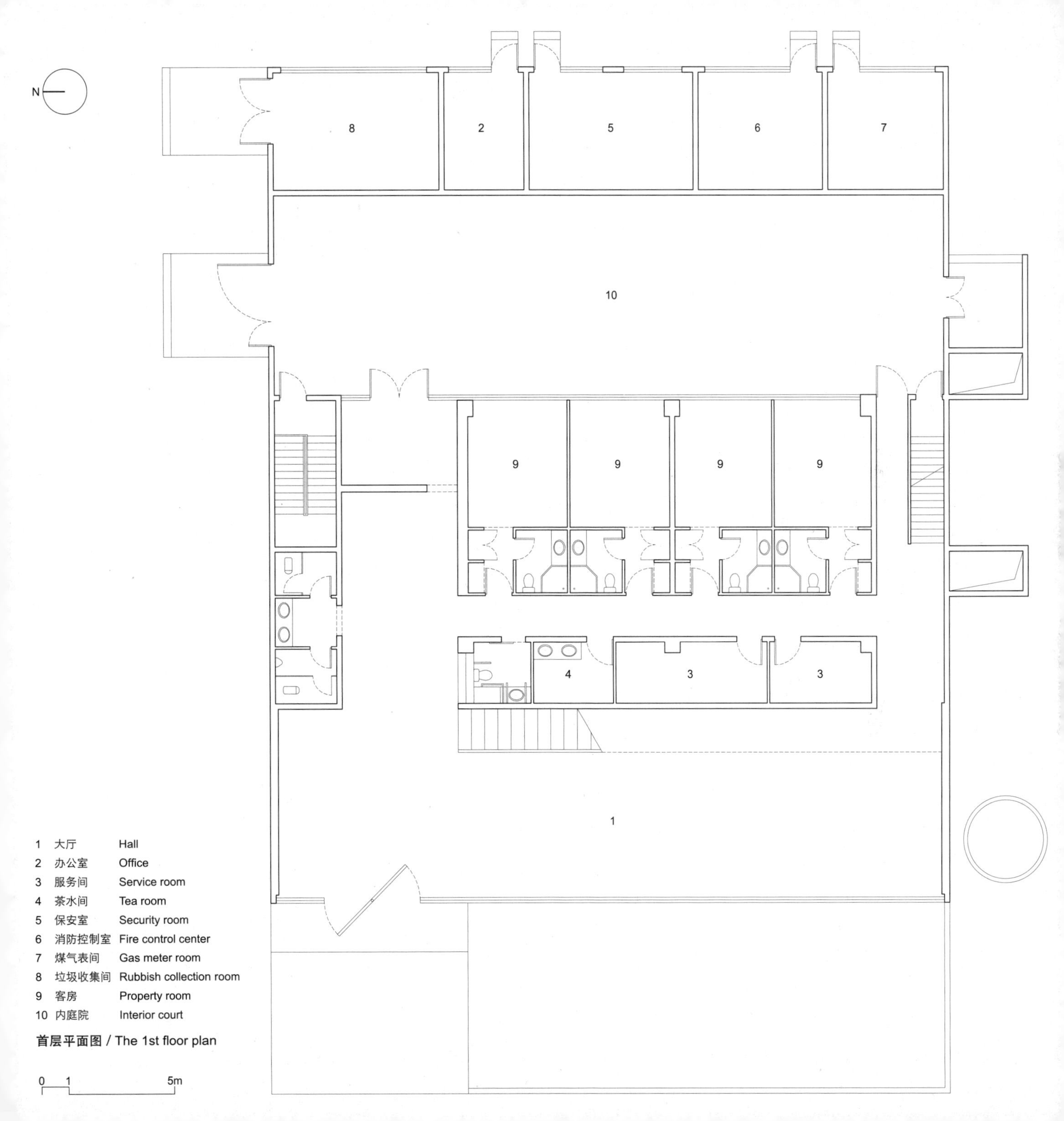


首层平面图 / The 1st floor plan

二层平面图 / The 2nd floor plan

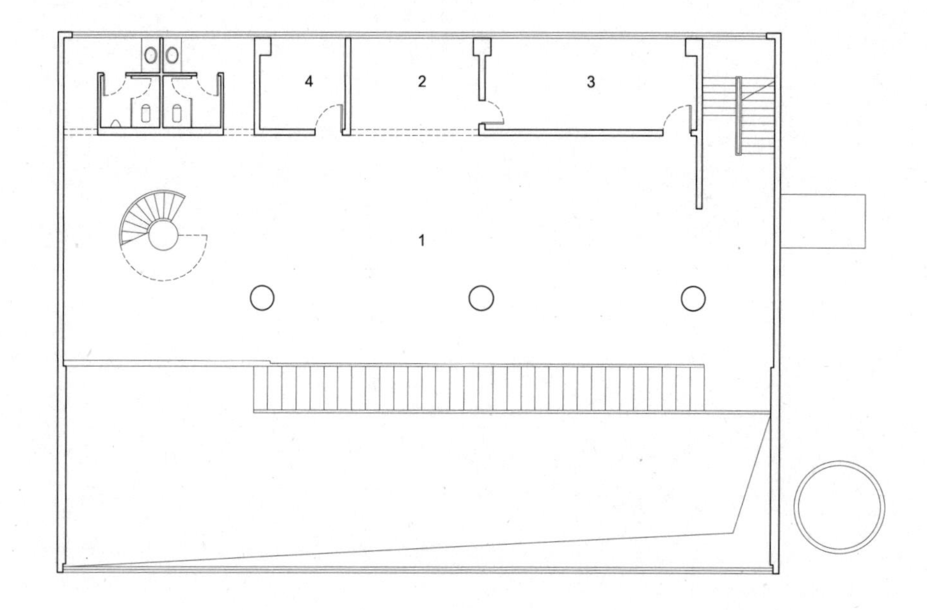


庐师山庄会所用地的西侧有一片大的中心绿地，使会所成为能让使用者体验中心庭院的装置所在是设计的最初动机。使用者希望会所的一层为咖啡厅并在一侧设置几间小的客房，二层为供居民们展示自己作品的画廊和举办一些交流性质活动的场所，三楼设置两个大的多功能会议室。设计时，考虑到西侧的风景，在一层大堂的一侧顺着景观的方向设计了一个长坡道，目的是让人不是停留在一个视点去看一个对象物，而是通过漫步并不断升高的过程，体会窗外风景的变化。

1 咖啡厅 Coffee hall
2 吧台 Bar
3 电加热间 Electric heating room
4 储藏室 Storeroom
5 会议室 Meeting room

三层平面图 / The 3rd floor plan

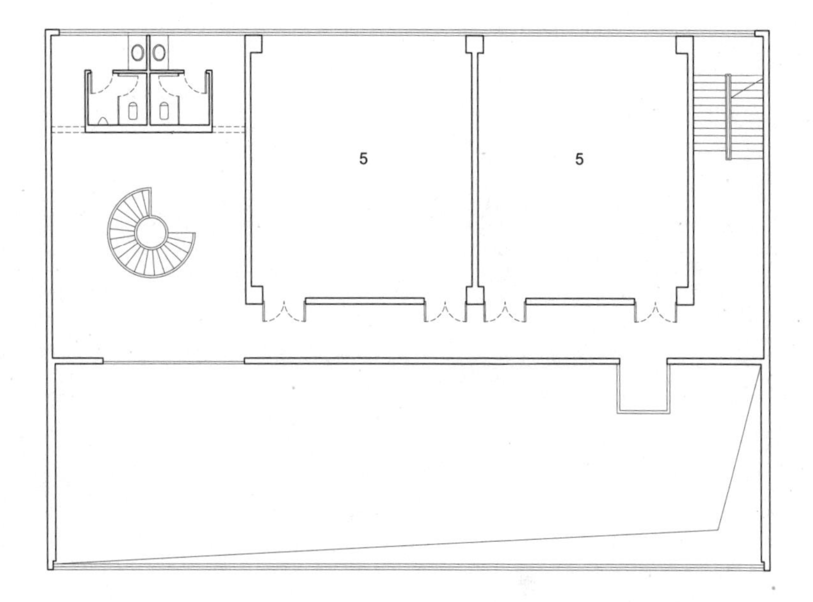


There is a large green space in the west side of Lushi Clubhouse's site. The design's primary intention is to make the users experience the installation in the atrium. The client required a cafe on the first floor, a gallery to exhibit residents' works and hold some activities, two big multi-function meeting rooms on the third floor; and several small guest rooms on one side of the first floor. A long sightseeing ramp is designed along the west side of lobby. The architectural promenade enables people seeing different views outside the window while escalating.

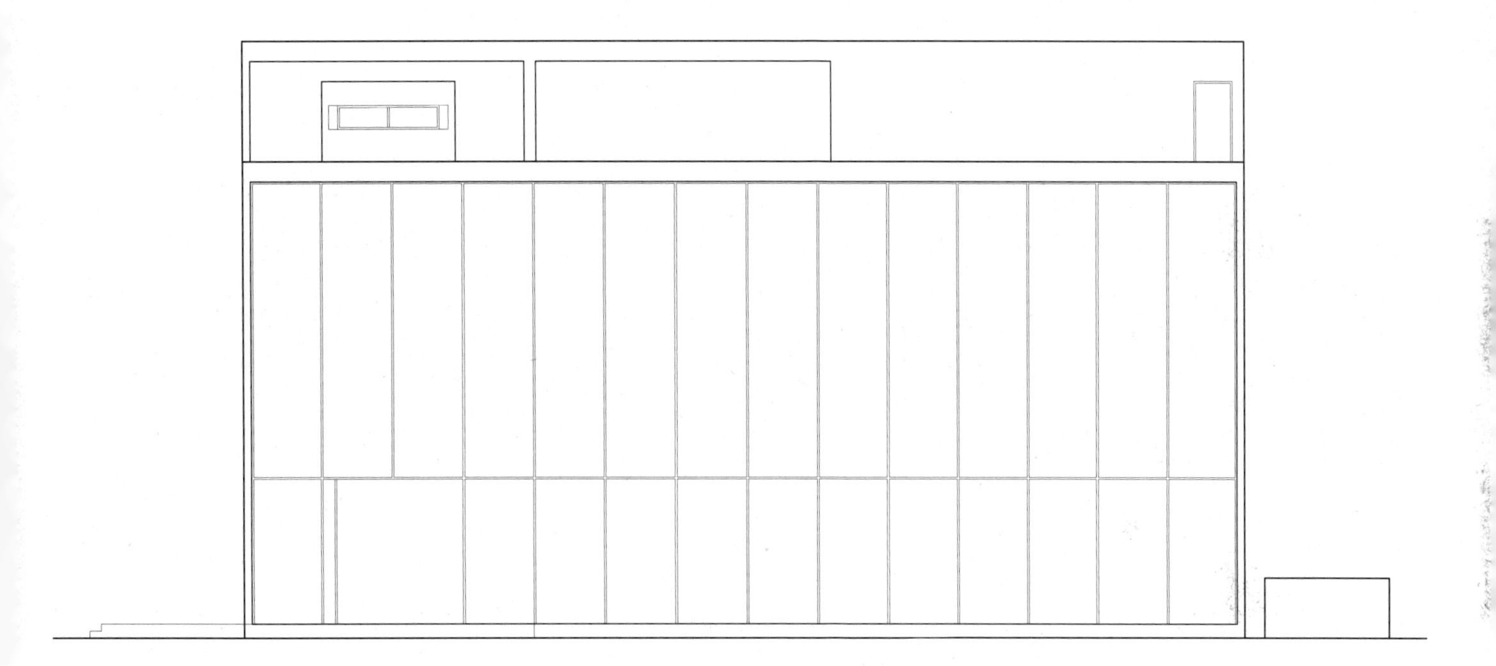

西立面图 / West elevation

在二楼一侧还有一个旋转楼梯通达三楼。这个旋转楼梯将从一楼到二楼的坡道过程中所观察到的平面性风景伴随这个通往三层的路径旋转进一步扭曲。旋转楼梯的上空是一个屋顶天窗。

There is a spiral stairway from 2nd to 3rd floor. It twists the plane ramp view from 1st to 2nd floor, at the same time itself becomes the visual centre of 2nd floor. Above the spiral stairway is a skylight.